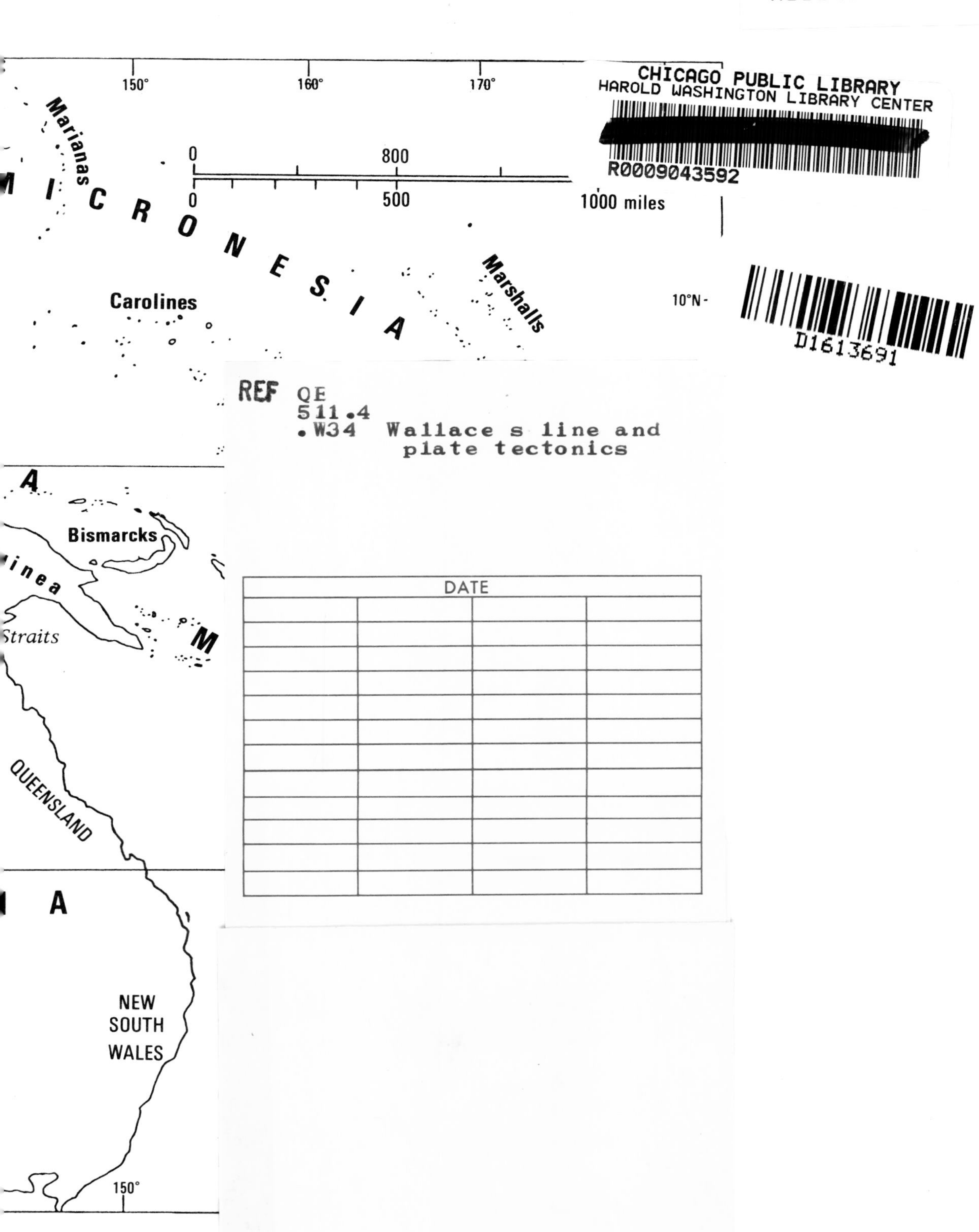
150°
160°
170°
Marianas
0
800
0
500
1000 miles
MICRONESIA
Marshalls
Carolines
10°N
A
Bismarcks
inea
Straits
M
QUEENSLAND
A
NEW
SOUTH
WALES
150°

OXFORD MONOGRAPHS ON BIOGEOGRAPHY

Editors: W. GEORGE, A. HALLAM, AND T. C. WHITMORE

OXFORD MONOGRAPHS ON BIOGEOGRAPHY

In an area of rapid change, this new series of Oxford monographs will reflect the impact on biogeographical studies of advanced techniques of data analysis. The subject is being revolutionized by radioisotope dating and pollen analysis, plate tectonics and population models, biochemical genetics and fossil ecology, cladistics and karyology, and spatial classification analyses. For both specialist and non-specialist, the Oxford Monographs on Biogeography will provide dynamic syntheses of the new developments.

WALLACE'S LINE AND PLATE TECTONICS

EDITED BY

T. C. WHITMORE
University Research Officer,
Commonwealth Forestry Institute,
Oxford

CLARENDON PRESS · OXFORD
1981

Oxford University Press, Walton Street, Oxford OX2 6DP
OXFORD LONDON GLASGOW
NEW YORK TORONTO MELBOURNE WELLINGTON
KUALA LUMPUR SINGAPORE JAKARTA HONG KONG TOKYO
DELHI BOMBAY CALCUTTA MADRAS KARACHI
NAIROBI DAR ES SALAAM CAPE TOWN

Published in the United States by Oxford
University Press, New York

British Library Cataloguing in Publication Data

Wallace's line and plate tectonics. – (Oxford
monographs on biogeography)
1. Plate tectonics
I. Whitmore, T. C.
551.1'36 QE511.4

ISBN 0-19-854545-2

Printed and bound in Great Britain at
The Camelot Press Ltd, Southampton

PREFACE

One of the sharpest zoogeographical divisions in the world divides the Malay archipelago in two. First brought to prominence by the famous naturalist A. R. Wallace in the mid-nineteenth century, Wallace's line, as it soon became named, has tantalized zoologists, and to a milder degree botanists, ever since.

In this book the new understanding of the geological history of the Malay archipelago achieved as a result of the theory of plate tectonics is described in some detail. The implications of this understanding for the interpretation of distribution patterns are then illustrated by consideration of vertebrate animals, palms, and several other plant groups.

It is hoped that the dramatic new discoveries in palaeogeography and their far-reaching implications for biology will interest the wide circle of students of biogeography. At the same time we hope to stimulate both botanists and zoologists to interpret their own research findings against the now considerable knowledge of the geological history of the region. For we think it can now be clearly demonstrated that the distributions of plants as well as of animals bear witness to this history. Thus the book is both for the generalist, and, at the same time, for the specialist in Malesian plants and animals. The conclusion is that, although a fresh look at distributions in the light of recent geophysical findings explains much, there still remain some distribution patterns which are very difficult to interpret. These are the challenges now to be faced.

Oxford
May 1981

T. C. W.

To

Professor C. G. G. J. van Steenis

Professor Emeritus, State University Leiden, Holland, and Director-General, Flora Malesiana Foundation. Mentor to many Malesian botanists.

ACKNOWLEDGEMENTS

It is not easy to integrate chapters by several authors to read as a coherent whole and the editor is most grateful to Mrs C. Taylor, Mrs P. Taylor, and Miss T. Hodgkinson for their forbearance in typing and retyping successive drafts on numerous occasions.

For help with Chapter 7 we acknowledge the kind assistance and criticism of Miss A. G. C. Grandison, Dr P. H. Greenwood, Dr R. F. Inger, Mr A. F. Stimson; and with Chapter 5 of Dr J. Flenley and Professor D. Walker. Mr P. A. Stott kindly read a draft of the whole book. Mr J. Lovett provided invaluable clerical assistance and Mrs J. Steele compiled the index to species.

The following acknowledgements are made for figures: Yayasan Indonesia Hijau and *Suara Alam* Figs. 7.1, 7.2, 7.3; M. J. E. Coode Fig. 8.12.

CONTENTS

CONTRIBUTORS

M. G. Audley-Charles, Department of Geological Sciences, Queen Mary College, University of London.

The Earl of Cranbrook, Great Glemham House, Great Glemham, Suffolk.

J. Dransfield, Royal Botanic Gardens, Kew.

W. George, Zoology Department, University of Oxford.

A. M. Hurley and **A. G. Smith**, Department of Earth Sciences, Sedgwick Museum, University of Cambridge.

T. C. Whitmore, Commonwealth Forestry Institute, University of Oxford.

NOTE ON PLACE NAMES AND UNITS

In general we follow *The Times atlas of the world, comprehensive edition* (5th edition, revised 1977) but have retained conventional English rendering for major localities, e.g. Celebes (Sulawesi). Place names mentioned in the text are shown on the endpaper map or the map of central Malesia (Fig. 4.9).

Banda Arc. The geological term for the double arc of islands from Flores east through Alor and Wetar and north to Banda (*Inner Banda Arc*) and from Raijua through Timor and the Tanimbar islands then north through the Kai islands and west to Ceram and Buru (*Outer Banda Arc*) (see Fig. 4.9).

Greater Sunda Islands. Borneo, Java, and Sumatra.

Inner Banda Arc. See Banda Arc.

Lesser Sunda Islands (Nusa Tenggara). The geographical term for the islands east of Java from Bali and Lombok eastwards to Damar and Babar (see Fig. 4.9).

Ma. The megayear, 1 000 000 years.

Malaysia. The political State comprising Peninsular Malaysia together with Sabah and Sarawak in northern Borneo.

Malesia. The biogeographical province stretching from Sumatra and the Malay peninsula south of the Kangar–Pattani line (Whitmore 1975) to the Bismarck archipelago.

Moluccas. The geographical term for the islands which occupy the region between Celebes (plus the Talaud and Sula islands, Butung and the Tukangbesi islands), the Lesser Sunda Islands (q.v.), Aru and New Guinea (plus Misool and Waigeo). The biggest Moluccan islands are Halmahera, Ceram, Buru, and Tanimbar (see Fig. 4.9).

Outer Banda Arc. See Banda Arc.

Papuasia. New Guinea, the Bismarck archipelago, and the Moluccas.

Sundaic. Pertaining to Sundaland.

Sundaland. The lands of the Sunda continental shelf, west of Wallace's line.

Sunda Shelf. See Sundaland.

Wallacea. The island region in central Malesia which lies between Wallace's line and Weber's line (Fig. 2.3).

West Gondwanaland. That part of Gondwanaland which today forms Africa and South America.

1 INTRODUCTION

T. C. Whitmore

The revolution in the earth sciences during the last two decades which has resulted from the theory of plate tectonics means that the palaeogeography of the globe during the Mesozoic and Cenozoic is now understood in general terms. Only details remain to be filled in.

This makes possible a profitable new interpretation of plant and animal distributions. For the first time the relative positions of land masses are known at different times in the past. The pattern is known with some certainty, although it is still very difficult to know where coastlines lay because only the stratigraphic record can tell us and this is very incomplete.

One of the classic and best known boundaries of zoogeography is the line, originally proposed in 1858 by A. R. Wallace, which runs through the middle of the Malay archipelago and marks the meeting of essentially Asian and Australian faunas (Chapter 2). The nature of this junction has been subject to much discussion and its exact position has been a matter of dispute. Wallace himself moved his line from west to east of Celebes late in his life, a decision to which new geological discoveries give a certain piquancy.

Until recently the various viewpoints on the palaeogeography and biogeography of the Malay archipelago had to stand or fall on their ability to fit all known facts into a plausible hypothesis. There was no external point of reference. Biologists tended to make geographical reconstructions; and geographers to use biological evidence. But now biologists can for the first time work with a broadly painted background of established palaeogeographical facts, albeit still lacking much fine detail, rather than mere suppositions. The former danger of circular argument, which has so often plagued discussions of biogeography, no longer befogs the scene.

Wegener, in his theory of continental drift, used the evidence of animal distributions in the Malay archipelago and epitomized by Wallace's line to support his view that south-east Asia and Australia had converged (Wegener 1924; and discussion in Hallam 1967). But it was to be nearly half a century before continental drift, reincarnated as plate tectonics and with the powerful evidence of sea-floor spreading as an essential component, became sufficiently well-established to provide this external point of reference for biogeographers.

The major events of global geology which affect the biogeography of the region of Wallace's line are the progressive break up of Gondwanaland from about 140 Ma and the drifting north of the Indian fragment to collide with Laurasia at about 55 Ma[1] (Chapter 3) and of the Australia/New Guinea fragment to collide with the south-east extremity of Laurasia at Celebes at only about 15 Ma, mid-Miocene (Chapters 3 and 4). The boundary between Gondwanaland and Laurasia lies within Celebes or just to its east (Fig. 4.7). Western Celebes is Laurasian, the Makassar Strait lies within Laurasia and has very probably been intermittently bridged from 15 Ma, the late Tertiary, onwards with more land above sea at its southern end. The Inner Banda island arc is Laurasian, the Outer Banda island arc and Moluccas are Gondwanic. Land connections from Laurasian Borneo to Gondwanic New Guinea were continuous or with only narrow sea gaps from some time between the late middle Miocene to late Pliocene (*c.* 12 Ma onwards), and from the Outer to the Inner Banda arc in the middle Pleistocene, 1 Ma, and probably latest Miocene/early Pliocene, 10 Ma, also.

The angiosperms, which first appear as fossils in the Lower Cretaceous, about 120 Ma, are commonly believed to have evolved considerably earlier than that, and somewhere in the region between Assam and Fiji, where the world's richest floras occur. The faunas there too are extremely rich and diverse. But we now see that, contrary to widespread belief, this great richness is not concomitant with long persisting conditions of stability. The new palaeogeographic understanding explodes the supposed cradle of the angiosperms.

The geological history outlined in Chapters 3 and 4 shows that plants and animals could have reached the Malay archipelago without crossing water from one of three sources: Laurasia, Gondwanaland via Australia, or Gondwanaland via India followed by south-eastwards migration. In addition, undoubtedly, some groups are autochthonous, having evolved more or less where they now live.

In this book we explore the present-day distributions of a few groups of plants and animals and interpret them in the light of this insight from palaeogeography. Much can be understood, but enigmas still remain.

[1] The megayear, see p. xii.

It is now known that during the Pleistocene there have been long periods when the climate of some parts of the world has been drier and more seasonal than today, and that at these times sea-level has been as much as 180 m lower. The two great present-day rain-forest blocks, one centred on Sumatra/Malaya/Borneo and the other on New Guinea, may have been smaller but there are no signs in the Malay archipelago of extensive desiccation (in contrast to both Africa and tropical America). Nevertheless, present-day distribution patterns of plants and animals are likely partly to reflect past differences in climate. The interpretation of range must penetrate this veil and take it, as well as palaeogeographical dispositions of the land, into account.

Our survey of animals focuses on vertebrates, in which Celebes is peculiarly isolated, though more Laurasian than Gondwanic in affinity, more so than islands to its north and south. Any line to be drawn in fact needs to circumscribe Celebes. The inference is that there have always been water barriers in central Malesia (Wallacea), which animal groups have traversed differently dependent on their dispersive ability.

Amongst plants the main investigation is of the palms, richly represented throughout the Malay archipelago. This is a large and complex family which recent study suggests encompasses five main lines of evolution and fifteen natural major alliances. Something is known about both the evolutionary history and the region of origin of the palms and they have been the subject of renewed intensive study in the Malay archipelago since 1965. The group is therefore ideal for an analysis of distribution patterns in the light of palaeogeography. The numerous species, genera, and natural major groups of palms in Malesia provide abundant information which can be set in a global context. It is commonplace that different groups of plants, both genera and families, have different ranges within the archipelago. In the concluding chapter we demonstrate that some of the commonly occurring distribution patterns can be explained in terms of the geographical history, though, as in the palms, many riddles still remain.

We end by discussing briefly one of the remaining major problems of global phytogeography, namely the occurrence of bihemispheric distribution patterns amongst flowering plants which approach or overlap in the Malay archipelago (Fig. 1.1) and have led to the suggestion, now seen to be improbable, that here lies the cradle of the group.

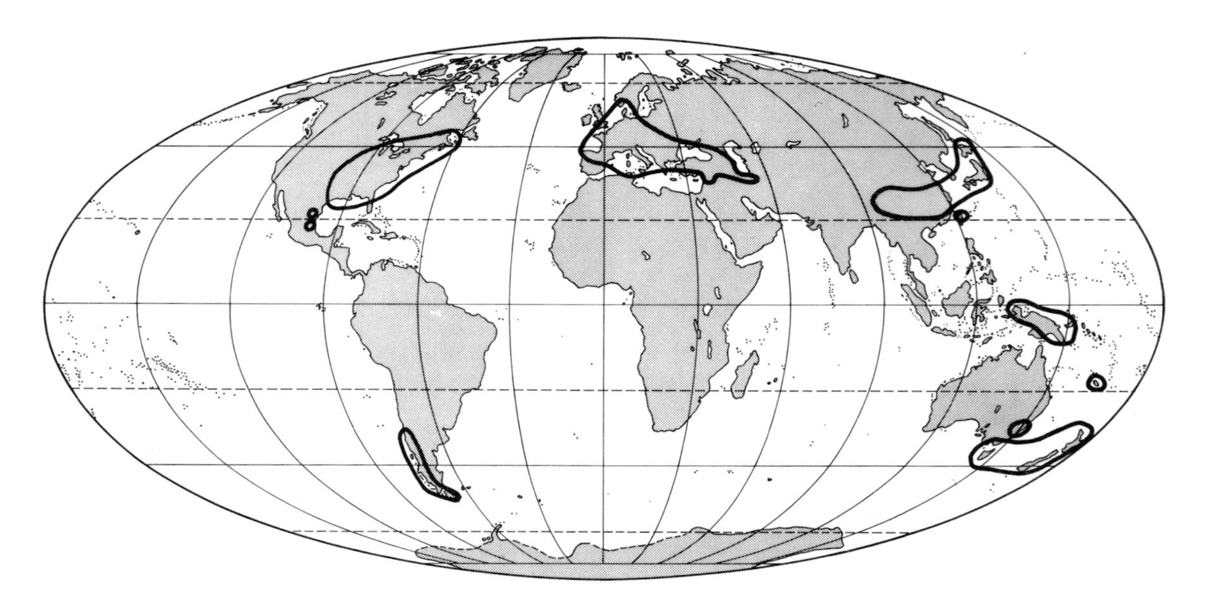

Fig. 1.1. The global distribution of Fagaceae subfamily Fagoideae, the beeches and southern beeches, has a bihemispheric pattern with ranges of the northern and southern genera most closely approaching each other in the Malay archipelago. Mollweide's elliptical equal area projection.

2 WALLACE AND HIS LINE

W. George

Wallace discovered the dramatic change in fauna which occurs in the middle of the Malay archipelago while he was still living there. His views on the exact position of the demarcation line changed with time. They were expounded in a letter to Bates in 1858; then in a series of publications which spanned the half century between 1859 and 1910. His discoveries have continued to tax the minds of zoogeographers who have proposed various refinements of the original discovery. Wallace realized the enigmatic nature of Celebes and, in searching for an interpretation, recognized the importance of an understanding of the palaeogeography of the region. His views and those of others provide a historical introduction to present-day understanding of both the palaeogeography and biogeography of the region.

The first indication that there were zoological regions, conforming to the major land masses of the world and brought into being by slowly evolving continents and slowly evolving animals, may be found in a letter that Alfred Russel Wallace sent from the island of Amboina in the eastern part of the Malay Archipelago in January 1858 to Henry Bates who was then back in London from his South American travels. 'In the Archipelago', he wrote, 'there are two distinct faunas rigidly circumscribed, which differ as much as those of South America and Africa, and more than those of Europe and North America. Yet there is nothing on the map or on the face of the islands to mark their limits. The boundary line often passes between islands closer than others in the same group. I believe the western part to be a separated portion of continental Asia, the eastern the fragmentary prolongation of a former pacific continent' (Marchant 1916).

Later that year, P. L. Sclater divided the world into six avifaunal regions. Unlike Wallace, he believed in stability and creation. But 'it is not yet possible', he wrote, 'to decide where the line runs which divides the Indian Zoology from the Australian' (Sclater 1858). Sclater included the Philippines, Borneo, Java, and Sumatra in the Indian region; and New Guinea, New Zealand, and some Pacific islands in the Australian region.

Wallace read Sclater's paper and congratulated him on his classification but proposed some emendations to boundaries between regions. In March 1859, Wallace argued that the boundary between the Indian and Australian 'zoology' should run between the islands of Bali and Lombok: because barbets reach Bali and not Lombok and cockatoos reach Lombok but not Bali. 'This, I think, settles that point.' Wallace then used parrots to run the boundary between Borneo and Celebes. He was amazed that two regions 'which have less in common than any other two upon the earth' should be separated by no major physical or climatic barrier. It seemed to confirm his belief that the western isles had once been part of Asia and that the eastern ones, including Celebes. Timor, the Moluccas and New Guinea, were remnants of a vast Pacific-Australian continent, but he had reservations about Celebes which he argued was peculiar and might represent fragments of a very ancient land which 'may have been connected at distant intervals with both regions' (Wallace 1859). Later that year, he described the ornithology of north Celebes and noted that it contained more Javanese species than Moluccan. He concluded that the island is 'one of the most interesting in the world to the philosophical ornithologist' (Wallace 1860*a*).

Wallace had been following Sclater in dividing the world into regions according to the distribution of birds but, in 1860, he decided that the world could be divided into regions that would 'hold good in every branch of zoology'. He sent a paper to England which was communicated to the Linnean Society by Charles Darwin (Wallace 1860*b*). In it he argued on the basis of his knowledge of mammals, birds, and insects for the division of the world into discrete faunal regions. The boundary between Bali and Lombok was retained in spite of his limited knowledge of the Bali fauna. The Moluccas were associated definitively with New Guinea and Australia. He had further reservations about Celebes which he attached to a vast Indo-Asian continent ('the last eastern fragment') and supposed that its main colonizations had been across the sea. 'Facts such as these', he wrote, 'can only be explained by a bold acceptance of vast changes in the surface of the earth.'

Wallace looked for an interpretation of faunal distribution in the light of past evolutionary events. But this new and dynamic approach was limited because he had only modern faunas from which to reconstruct the past. No other line of reasoning was possible because there was nothing known of either the palaeogeography or the palaeontology of the Malay archipelago and no obvious physical differences between the islands.

However, his account of a visit by his assistant Charles Allen to the Sula Islands (Fig. 4.9), reported to the Zoological Society of London in 1862, showed how effective this approach was (Wallace 1862). The Sula Islands had a mixture of Moluccan and Celebesian forms which, because of their geographical position, did not surprise Wallace. But he noticed that there were far more Celebesian components of the fauna than Moluccan: according to his calculations, nearly double. So he concluded that the Sula Islands are an outlying fragment of Celebes and must have had a connection with that island in the past.

On further consideration, Wallace later associated the island of Buru with the Sula Islands: the presence of such animals as the babirusa pig (Fig. 7.2) and the megapode brush turkeys made it likely that these islands once had a connection (Wallace 1869).

Back in London, Wallace read a paper to the Royal Geographical Society in 1863 on the geography of the Malay Archipelago. This was mainly deduced from zoogeographical distributions but he also considered the depth of the oceans round the islands, the distribution of volcanoes and coral islands in the archipelago which had made it an unstable area for long periods of time, the vegetation of the islands and their climate. Wallace argued that it was an area of change, of evolution.

The 100 fathom (180 m) line running between Borneo and Celebes made Borneo part of an Asian continent. On similar evidence, New Guinea and the Aru Islands were part of an Australian continent (Fig. 2.1) as G. W. Earl(e) had already noticed in 1845. Zoological evidence was used to put the intermediate islands in one region or the other. Celebes and all islands to the east, including the outer Banda Arc islands, were therefore put in the Australian region. The Philippines, though deficient in many striking Asian mammals such as gibbons and rhinoceroses, were part of an Asian continent. In the south, the line ran between Bali and Lombok.

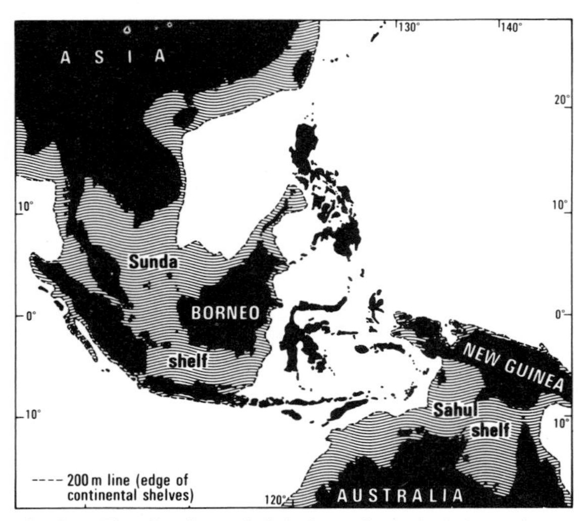

Fig. 2.1. The Sunda and Sahul continental shelves. (From Whitmore 1975.)

Moreover, in this paper Wallace had changed his mind about Celebes: in 1860 Celebes was the outlying eastern fragment of an Indo-Asian continent, in 1863 it was the farthest westward extension of an Australian continent. This revision was based on the remarkable differences Wallace had observed and later substantiated between the faunas of Borneo and Celebes, in spite of their geographical closeness: differences in mammals (no hedgehogs and only viverrid carnivores on Celebes, for example), differences in birds (no trogons or barbets but megapodes on Celebes), differences in insects (green–gold *Pachyrhynchus* weevils absent from Borneo) (Wallace 1863, 1869, 1880) (Figs 2.2 and 2.3).

In an 1865 article on the pigeons of the Malay Archipelago he confirmed that Celebes had associations through the Sula Islands with the Moluccas and New Guinea. He recognized a centre of pigeon diversification on New Guinea which extended outwards to include the Solomon Islands to the east and the Moluccas and Timor to the west. For example, the range from the Pacific islands to north Celebes of the brown-backed ground pigeon *Chalcophaps stephani* put Celebes into the eastern subregion of the archipelago. But there were other pigeons on Celebes, like the green pigeons *Treron*, whose affinities were with the western islands. 'Celebes', he decided, 'is a very isolated and remarkable island, which, from the variety and peculiarity of its productions, appears to be the remnant of some extensive land, which existed anterior to the present distribution of land and water in the surrounding regions' (Wallace 1865).

In his 1863 paper Wallace drew a red line on the map passing down the Makassar Strait. To the west, he wrote 'Indo-Malayan region' and, to the east, he wrote 'Austro-Malayan region'. This became 'Wallace's line' in 1868 when T. H. Huxley in a paper on gallo-columbine birds referred to a 'boundary in question'. It was supposed to 'coincide with what may be called "Wallace's line" between the Indian and Papuan divisions of the Malay Archipelago'. It did not 'coincide' in the north for in his own figure Huxley's line passed to the west of the Philippines but whether this was intentional is not clear from the text. Whatever the intention, it resulted in *two* Wallace's lines: one to the east of the Philippines and one to their west (Fig. 2.4).

Later, Wallace reproduced his 1863 line in his great

work *The geographical distribution of animals*: it still ran to the east of the Philippines (Wallace 1876).

Wallace had put Celebes in the 'Austro-Malayan region' in 1863 and 1876 but he argued that, if the species and genera which are common to Celebes and the surrounding islands are considered, 'we must admit that the connexion seems rather with the Oriental than with the Australian region'. However, he also argued that, if the proportion of species and genera present to those absent is considered, 'we seem justified in stating that the Austro-Malay element is rather the most fully represented'. From this, Wallace concluded that Celebes had probably never been united by extensive land to either the east or the west (Wallace 1876).

By the time *Island life* appeared in 1880, Celebes had become an 'anomalous island' along with New Zealand. Its poor mammal fauna, although much of it of Asian derivation, made it unlikely that the island had formed part of an Asian continent for more than a short period of time. Or perhaps, he argued, it had been no more closely linked than by a string of islands. But 'the question at issue can only be finally determined by geological investigations'. To Wallace, the bird fauna of Celebes appeared more Australian than to later investigators but the island, 'both by what it has and what it wants, occupies such an exactly intermediate position between the Oriental and Australian regions that it will perhaps ever remain a mere matter of opinion with which it should properly be associated' (Wallace 1880).

Wallace's last word on Celebes came in 1910 when he decided that his famous line should be redrawn: 'I came to the conclusion that Celebes was really an outlier of the Asiatic continent but separated at a much earlier date, and that therefore Wallace's line must be drawn east of Celebes and the Philippines' (Wallace 1910).

By the early twentieth century, other lines, shown on Fig. 2.4, had appeared among the islands based on the distribution of one group of animals or another (George 1964; Simpson 1977). Other than Wallace's, the only one that has persisted is Weber's line of 1904. This line was based on the distribution of molluscs and mammals and conformed, more or less, to the 100 fathom (180 m) sea line in the east of the archipelago.

Thus, two shallow water lines had been recognized (Fig. 2.1) and many authors maintain that, as these lines delineated ancient continents, it was possible that there had always been only ocean and islands in between. Consequently, each island had obtained a characteristically peculiar and unique fauna. For this reason, Dickerson (1928) proposed that the island area should be considered as a separate region (a suggestion already made by Wallace in 1863) and gave it the name Wallacea.

Wallace's line has persisted amongst zoologists as an important concept. It has much less utility to botanists. There is a disjunction at the Makassar Strait, with 297 genera of flowering plants reaching their eastern limit there, but the boundary between Java plus the Lesser Sunda Islands and the rest of the archipelago is of equal importance (van Steenis 1950). There are approximately 2300 genera of flowering plants in total in the archipelago and for the great majority Wallace's line is unimportant.

Some zoologists, like Raven in 1935, considered the validity of Wallace's line on the basis of the proportion of mammals that had crossed the line going east compared with those that had not and came to the conclusion that Wallace's line marked a boundary which was the eastern limit of the great majority of East Indian mammals, like rhinoceroses and elephants. Others made their assessment on the proportion of western and eastern elements to be found on each island in Wallacea. Thus, Rensch in 1936, following Mertens (1934), calculated that 88 per cent of Celebes reptiles, 80 per cent of the amphibia and 88 per cent of the butterflies were of western origin which was a similar proportion to that found on Lombok and more than twice as high as for the Kai Islands. Following the same line of argument for Australo-Malayan birds, Ernst Mayr calculated that 67.6 per cent of the passerines were from the west and decided that 'there is no doubt, Celebes must be included with the Oriental region' (Mayr 1944).

After a detailed examination of the lines and of Wallacea, Mayr along with several authors, at a Linnean Society discussion (Scrivenor *et al.* 1943), concluded that Weber's line was a better boundary between the faunas of the two great regions than Wallace's line although the butterflies, for example, left Celebes as anomalous as ever (Corbet 1943). Thus, the conclusion was reached that most of the modern fauna in Wallacea consists of more Asian types than Australian.

Other authors have taken an ecological approach to boundaries. Already, in 1846, Salomon Müller had drawn a line very similar to Wallace's but which he determined ecologically pointing out, for example, that arid conditions run eastwards from Java to Tanimbar and affect the vegetation. In 1857, Zollinger ran a similar floristic line from east of Timor to east of Celebes. And Wallace pointed out, in 1863, that the islands could be classified according to whether they were forested or not, those to the west of his line together with Celebes and New Guinea having essentially forest biomes and the rest more arid conditions. Lincoln (1975), studying the birds of Bali and Lombok, came to the conclusion that Wallace's line marks the division between a rich continental fauna associated with high rainfall, forests and varied habitats and an impoverished fauna associated with low

Fig. 2.2. Wallace's selection of Sundaic animals to contrast with Papuasian animals (left to right and top to bottom): the western tarsier *Tarsius bancanus*, one flying lemur *Cynocephalus variegatus* in flight and another one seated, the pentail treeshrew *Ptilocerus lowii*, the Malay tapir *Tapirus indicus* and a couple of lesser mousedeer *Tragulus javanicus*.

PLATE X.

SCENE IN NEW GUINEA, WITH CHARACTERISTIC ANIMALS.

Fig. 2.3. Wallace's selection of Papuasian animals to contrast with the Sundaic animals (left to right and top to bottom): a tree kangaroo *Dendrolagus inustis*, the fairy lory *Charmosyna papou*, the twelve-wired bird of paradise *Seleucides melanoleuca*, the common paradise kingfisher *Tanysiptera galatea*, and a crowned pigeon *Goura cristata* (*coronata*) (Figs 2.2 and 2.3: Wallace (1876), Vol. 1).

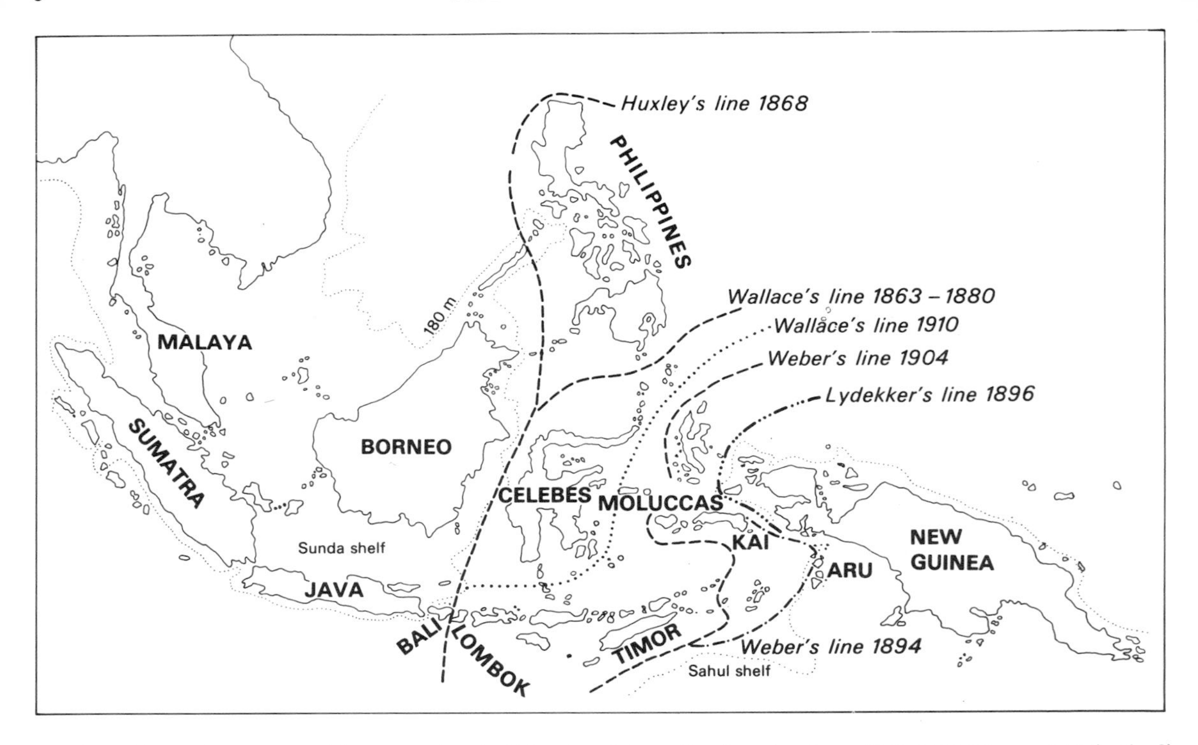

Fig. 2.4 Different lines suggested for separating the Oriental and Australian faunal regions, 1868–1910 (George (1964), Fig. 3).

rainfall, thorn scrub and restricted habitats. More recently, van Steenis (1979) suggested that climatic factors, both past and present, have been of paramount importance in preventing the spread of Australian plants. Adapted to poor soil conditions and an arid climate, they soon reached their western limit.

None of these analyses takes into account the past and many biogeographers would maintain that this is correct: regional faunas should be defined in terms of what they are, not what they were (for example de Lattin (1967) and Durden (1974)).

But the two shallow water lines are there and some zoogeographers, like the first of all evolutionary zoogeographers Wallace himself, believe that regional zoogeography should reflect the geological, floral, and faunal history of the area as well as present day distribution. To this end, several Dutch expeditions worked over the area in the 1930s (Umbgrove 1949) and were able to show the existence of an Asian continent embracing the islands as far east as Borneo and Java and an Australian continent embracing the south of New Guinea but leaving most of the intervening islands either submerged or as discrete islands during the whole of the Tertiary (van Bemmelen 1949).

Plate tectonics and the theory of continental drift have now shed new light on the geological history of the Malay archipelago. And, in spite of the objections of the strict faunal analysis school, it could be concluded with Wallace that 'it is true that we here reach the extremist limits of speculation; but when we have before us such singular phenomena as are presented by the fauna of the island of Celebes, we can hardly help endeavouring to picture to our imagination by what past changes of land and sea (in themselves not improbable) the actual conditions of things may have been brought about' (Wallace 1876).

3 CONTINENTAL MOVEMENTS IN THE MESOZOIC AND CENOZOIC

M. G. Audley-Charles, A. M. Hurley, and A. G. Smith

A series of maps (Figs 3.1 to 3.10) drawn at 20-Ma intervals from 200 Ma (latest Triassic time) to the present day show the fragmentation and dispersal of the supercontinent of Gondwanaland into the fragments known today: South America (only partly appearing on the maps); Africa; Arabia; Madagascar; India; Australia; New Zealand; and Antarctica. Africa, Arabia, India, Australia/New Guinea and New Zealand migrate northwards; between Africa and Arabia to the south and western Eurasia to the north lies a complex collision zone; to the east India collides with central Asia; and Australia (with New Guinea) collides with south-east Asia in the region of Wallace's line. These major Mesozoic and Cenozoic continental movements have taken place contemporaneously with the rise of mammals and flowering plants. They provide the backdrop against which to interpret the present-day biogeography of the Malay Archipelago.

HOW THE PALAEOCONTINENTAL MAPS WERE MADE

The maps in this chapter were made by drawing global reconstructions based on geophysical data. The phases of continental separation can be traced from the striped pattern of magnetization on the Atlantic and Indian Ocean floor. As Gondwanaland split apart, new ocean floor welled up from the earth's mantle at the mid-ocean ridge crests. This hot material cooled, acquired the magnetization appropriate to the time, and moved symmetrically outward from both sides of the ridge. The earth's magnetic field changes its polarity at irregular intervals averaging about 1 Ma. The magnetism of the ocean floor rocks can be measured by ships at the surface. The reversal pattern gives a characteristic striping to a map of their direction of magnetization. The positions of the change of direction of magnetization are known as anomalies, and these are each assigned a number for purposes of identification. Because the reversal pattern varies in a known manner with time, the age of the ocean floor can be determined from a magnetic survey.

The earth's magnetic field averages to that of a dipole—the field made by a bar magnet—at the earth's centre with its axis parallel to the earth's spin axis. There is a simple relationship between the angle at which the lines of force from such a dipole emerge at the earth's surface and the geographic latitude. The direction of the ancient magnetic field is preserved in ancient sediments or lava flows. Thus, by averaging a sufficiently large number of ancient field directions, the ancient magnetic poles can be determined and hence the former position of the observation site can be estimated in past time. A map of the continents in their past positions can be made by using the ocean-floor data to reposition the continents relative to one another and then employing the ancient field direction to orient the continental reassembly so that it has the correct palaeolatitudes.

Global reconstructions have been made in two stages. Initially a continent is chosen as a reference (in this case Africa). For any past period the continents around Africa are moved towards it according to the published ocean-floor spreading data, to produce a continental reassembly. In the second stage the reassembly is projected as a map, with a palaeogeographic grid superposed.

First, a list of well-dated magnetic anomalies is made so that those in the chief oceans can be matched with each other as well as possible. The best fit of the anomalies in the ocean between two continents can be described by a rotation about an axis through the earth's centre. The point where the axis cuts its surface is known as the tectonic rotation pole. For example, the continental positions at 40 Ma are calculated as follows. South America can be moved to Africa so that anomalies number 13 (about 36 Ma) either side of the mid-Atlantic ridge in the South Atlantic, match as well as possible. To obtain the value of the 40 Ma rotation of Africa to South America, the rotations necessary to match anomalies numbers 5 (9 Ma) and 13 (36 Ma) are summed. Then because the next well dated anomaly is number 34 (80 Ma) the appropriate fraction of the 36 to 80 Ma incremental rotation is added to the previous sum. Similarly, North America can be moved to Africa by fitting together the 36 Ma anomalies in the central Atlantic, and then by interpolation the total 40 Ma rotation can be calculated. In the same way a set of rotations may be determined to reassemble all the continents relative to Africa, at any chosen time. These rotations are determined automatically by a computer program that looks up the appropriate values in a file listing all known published incremental rotation data.

The 40 Ma reassembly is converted to a geographic map by estimating the position of the past geographic poles from palaeomagnetic poles on the continents. This is done by another computer program which automatically examines all known palaeomagnetic measurements on the stable parts of the reassembly, and selects those within the appropriate time range (in this example 40 ± 10 Ma). These palaeomagnetic poles are rotated from their present-day positions to their positions on the reassembly and their average is taken as the geographic pole at 40 Ma.

For plotting plant and animal distributions an equal area map is the most advantageous. Here Lambert's equal area projection is used, a projection that also has the ability to show both geographic poles on one map.

These maps are computer drawn. For convenience only, all the continents are shown with their present coastlines and 500 fathom (or 1000-metre) submarine contours and their present geographic grid projected on to the map at 10° intervals. This enables data to be plotted in their appropriate positions on the map.

Wallace's line lies at about 120 degrees east; therefore the map centres are chosen to be at 120° east on the equator. For the present day there is no ambiguity in the longitude. The '120-degree longitude' on all maps older than the present day map is arbitrary. However, for the purposes of relating biological changes to continental movements it is necessary only to know the former *relative* positions of the continents and their former latitudes – data that are on the maps.

Geologically uncorrected maps showing the whole world have been made by similar methods (Smith and Briden 1977). Further explanations of the way in which the maps have been made are given in Smith, Hurley, and Briden (in press).

The uncertainties in the maps are threefold: errors in the rotations; errors in the mean palaeomagnetic pole position, and errors in the projected shapes. It is difficult to quantify any of these errors. The greatest uncertainty is probably the shape and position of all those areas that have been deformed.

However, there are five effects that cannot be corrected but might be of importance to the biological evolution of the area. These are, first, the geological difficulties of recognizing sutures (i.e. places where continents have collided). Secondly, it is possible that some continental fragments have vanished, carried down into the earth's interior by tectonic processes associated with collision. Thirdly, small continental fragments may have acted as rafts across the Tethys. Fourthly, there is the possibility that transient land bridges existed in the Indian Ocean. Finally, it is difficult to unscramble the original geometry in those areas affected by deformation.

Sutures

When continents collide they create an orogenic belt—a belt of deformed rock—which marks the zone where they have been joined together. Such belts may be up to several hundred kilometres wide. The represent the scars where two continents have been spliced together by collision. On Figs 3.1 to 3.10, possible sutures including a Palaeozoic suture (the Urals), are shown by lines of crosses. Unfortunately, in many cases the suture is not well defined, and some parts assigned to Gondwanaland may have belonged to Eurasia and vice versa, but the areas involved are unlikely to be large.

Vanished continental fragments

Many geologists believe that the Tibetan plateau stands high because part of the northern Indian continent has been overridden by Tibet. If so, when India is pulled away from Tibet to make a reconstruction of the past geography, one should pull out that piece of crust now overridden by Tibet, adding it to the northern edge of India. This has not been done on the figures. For this reason, although geologically the collision between India and Tibet is believed to have taken place at about 55 Ma, the maps appear to show an oceanic region between India and Tibet at this time.

Transient oceanic land bridges

When continents were regarded as fixed, transient land bridges were postulated to explain how animals and plants had migrated from one continent to another. The 5000 km long Ninetyeast Ridge in the Indian Ocean has at times been regarded as such a bridge.

At present, this feature is wholly submarine; its crest deepens from south to north and it becomes older in the same direction (Kemp and Harris 1975). As soon as it has been created by spreading, all ocean-floor starts to cool down. As it does so, it contracts and the overlying water depth gradually increases at a predictable rate. Thus whatever its origin, the Ninetyeast Ridge was nearer the surface in past time. Dredging shows that in later Cretaceous and early Tertiary time (about 100 to 50 Ma ago) its crest was marked by a line of small islands that were submerged by Miocene time, about 20 Ma ago (McGowran 1978). Kemp and Harris (1975) analysed the pollen found in sediments dredged from the ridge and found a pronounced similarity with early Tertiary floras of Australia and New Zealand, including wind-borne conifer pollen. They concluded that the islands were colonized by plants capable of migration across water. For even though the ridge was closer to continents when

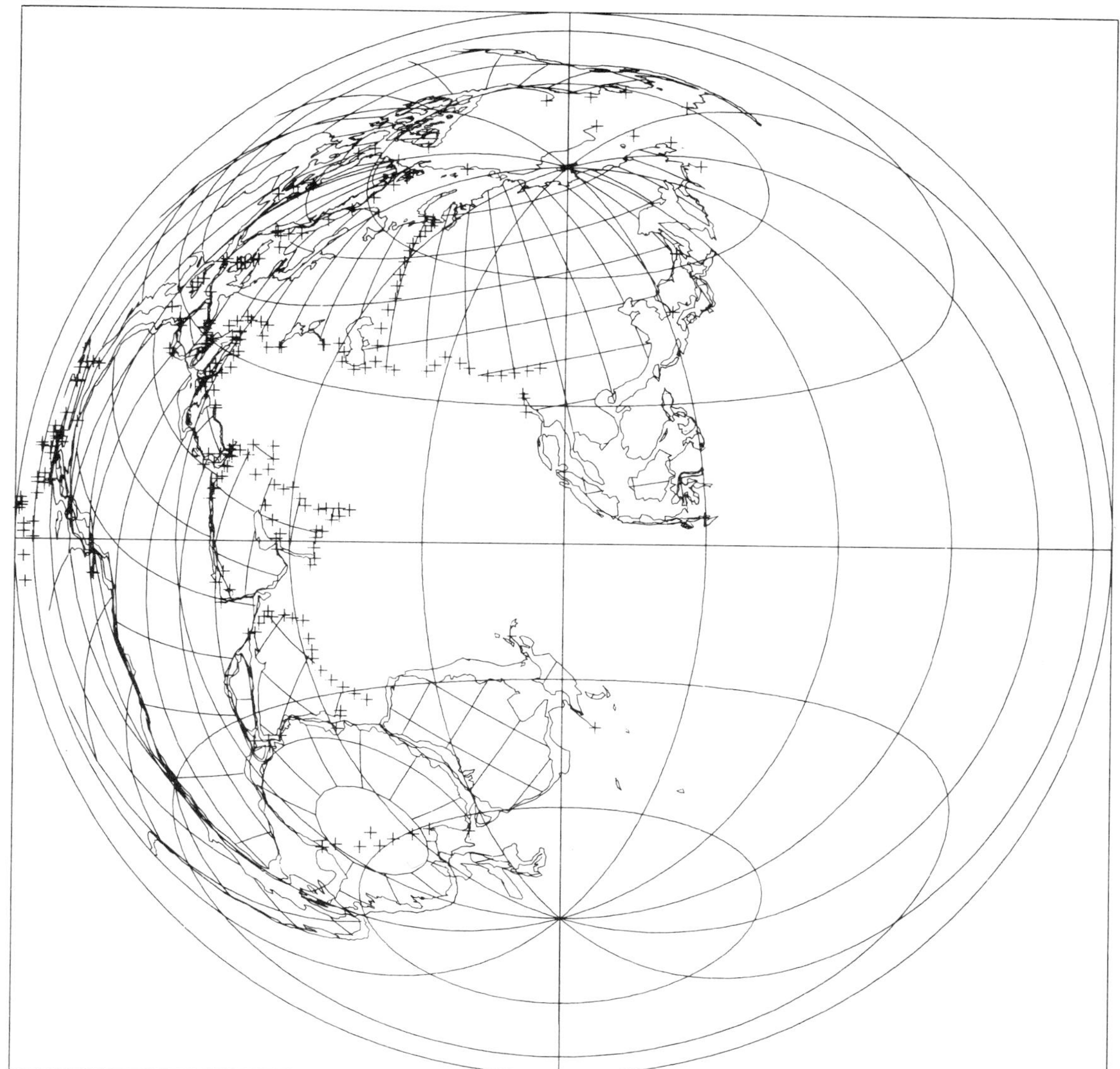

Fig. 3.1. The world at 200 Ma. This and Figs 3.2–3.10 are Lambert equal area projection centred on the Equator and 120°E. Present-day coastlines and 500 fathom (1000 m) submarine contours shown. See text.

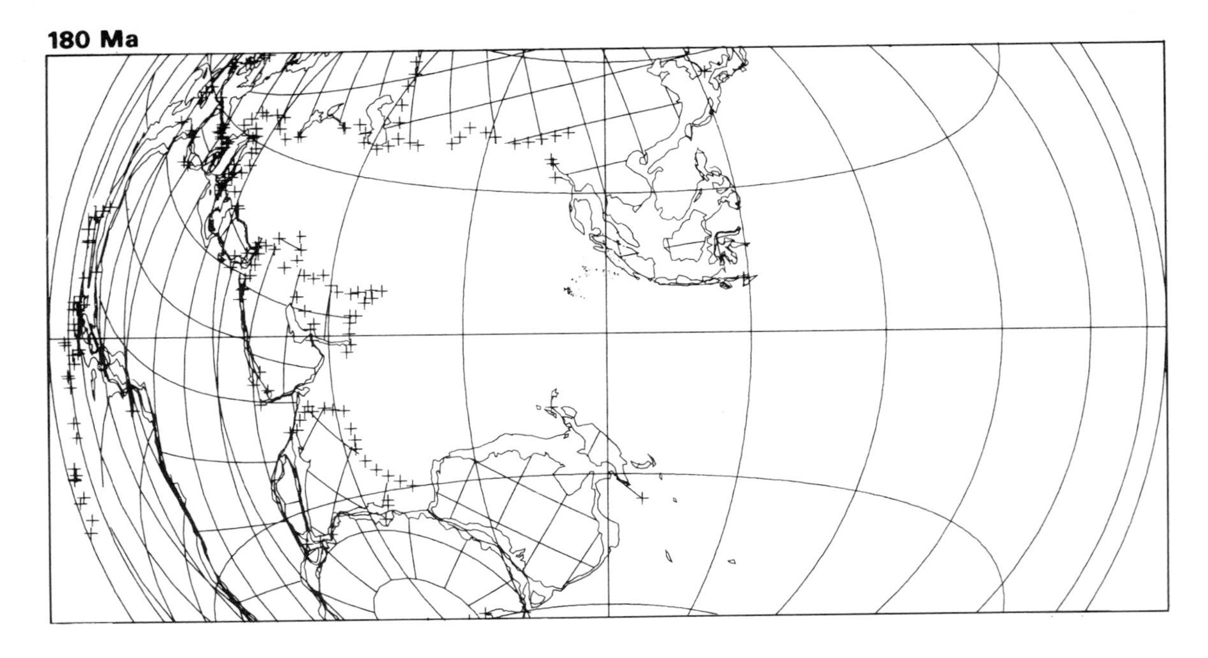

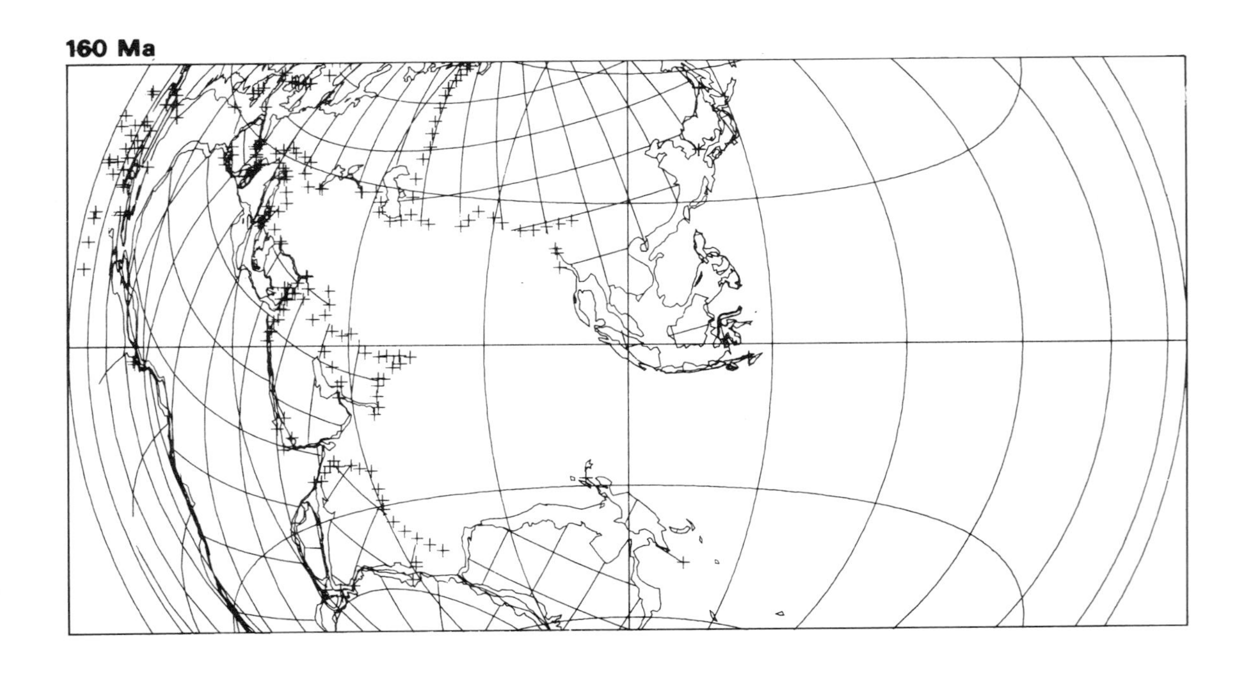

Fig. 3.2. The 'old world' tropics and Southern Hemisphere at 180 and 160 Ma. This is a period with little change but Africa starts to separate from America at about 180 Ma.

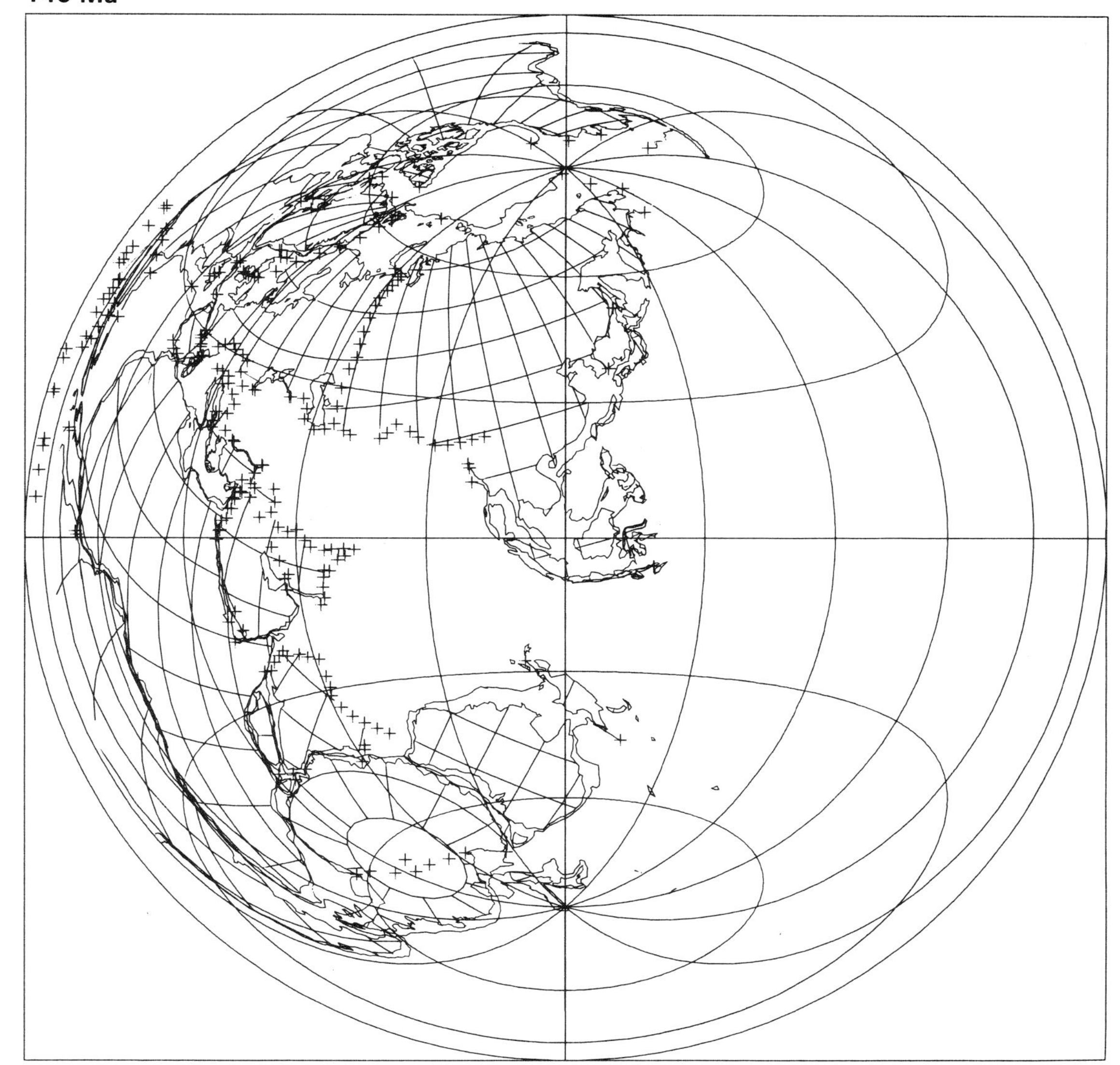

Fig. 3.3. The world at 140 Ma when India–Madagascar probably splits from Antarctica–Australia/New Guinea and from Africa.

120 Ma

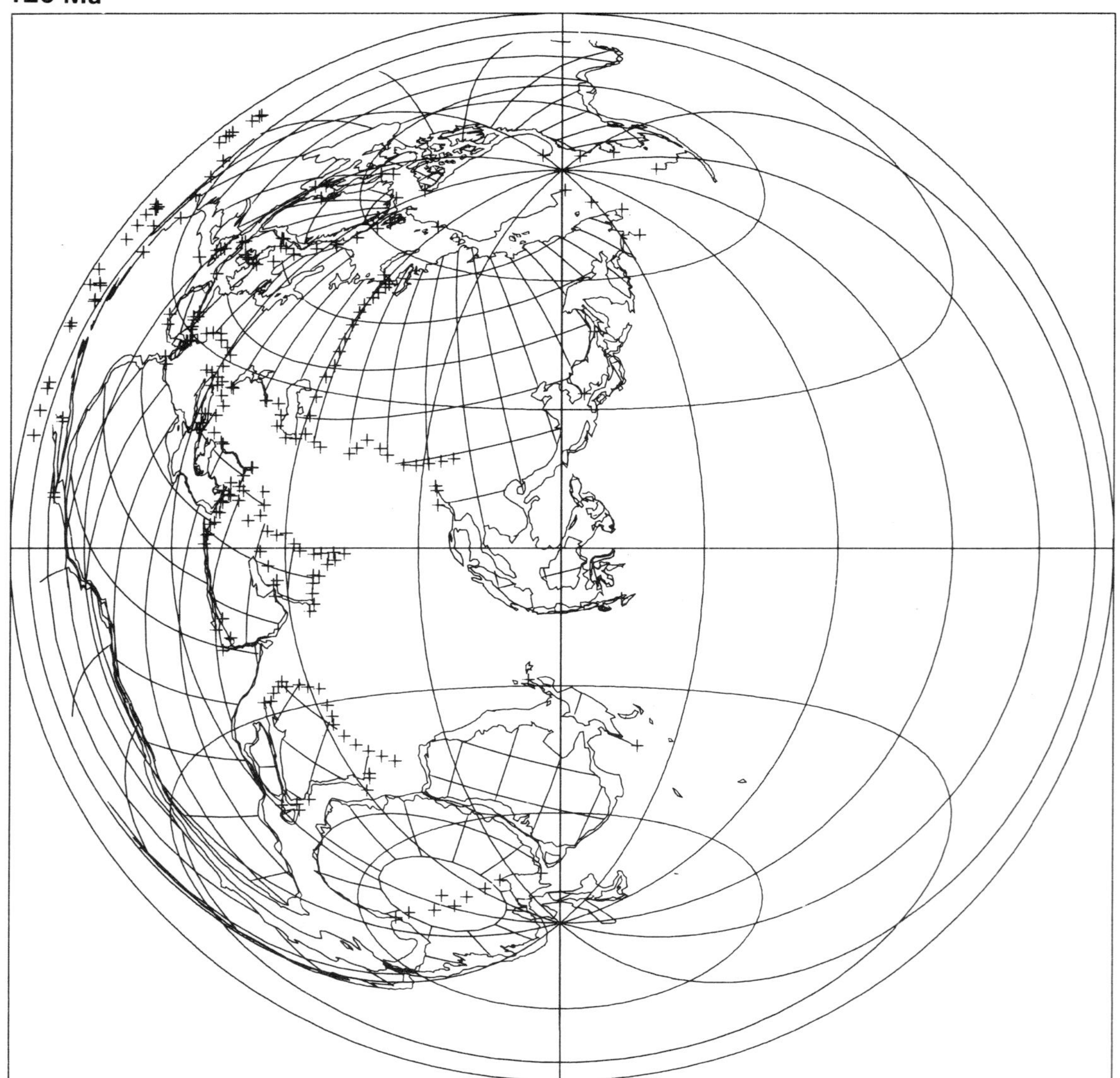

Fig. 3.4. The world at 120 Ma. India–Madagascar has probably now separated from Antarctica–Australia/New Guinea and America from Africa. South America probably starts to break away from Africa.

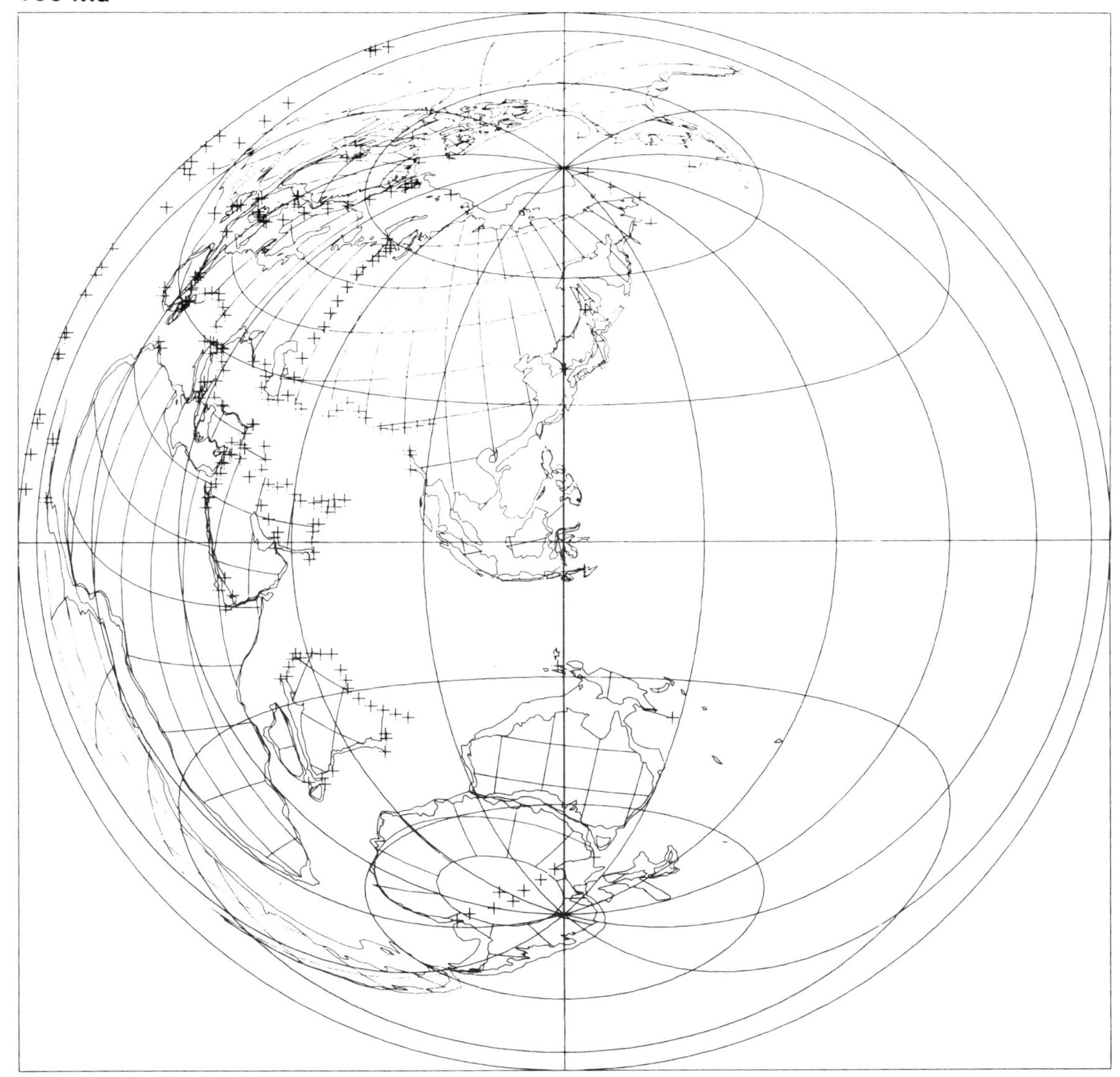

Fig. 3.5. The world at 100 Ma. India has probably now separated from Madagascar and started to drift rapidly northwards. America and Africa continue to move apart.

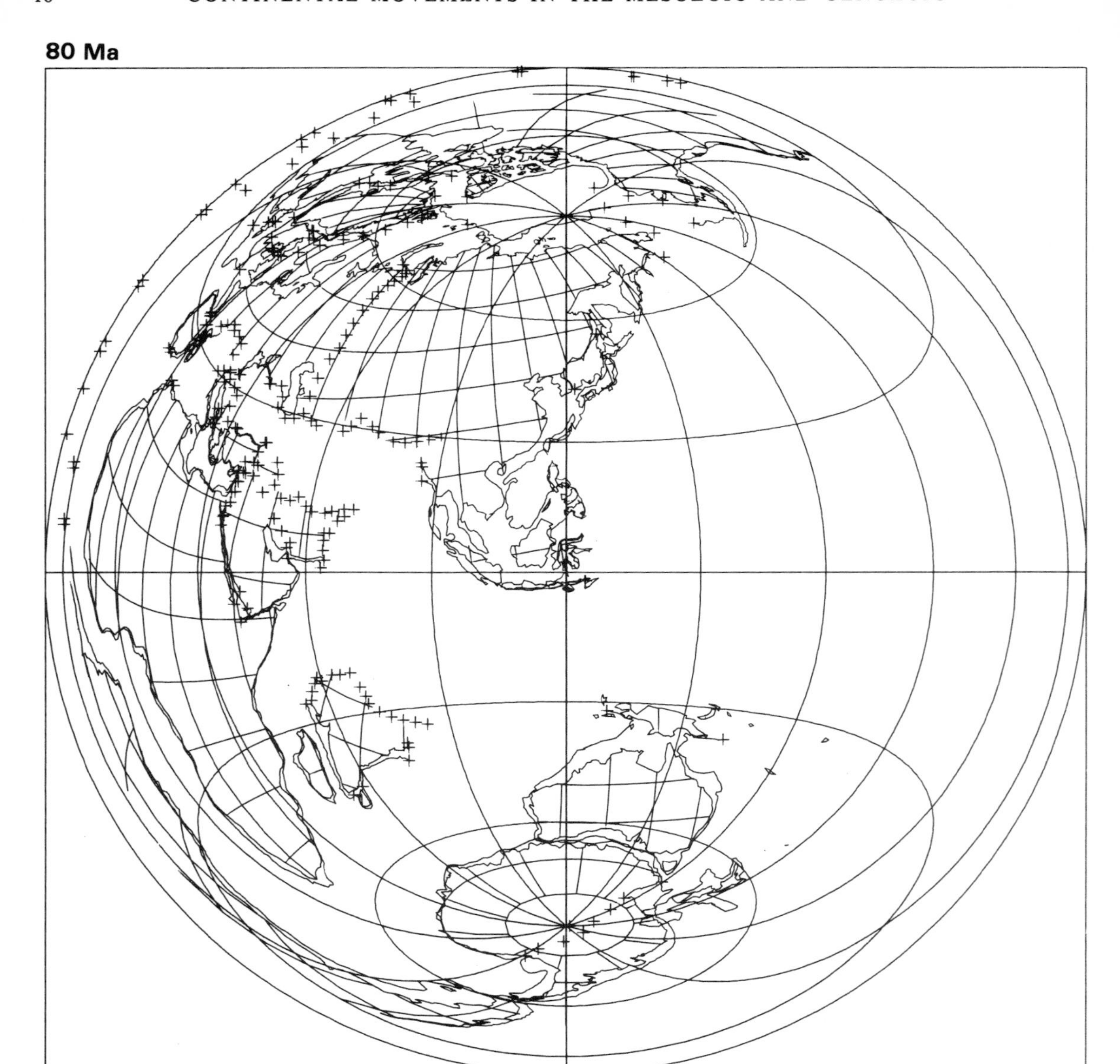

Fig. 3.6 The world at 80 Ma. The same movements continue and collisions have probably begun to take place in the Mediterranean region.

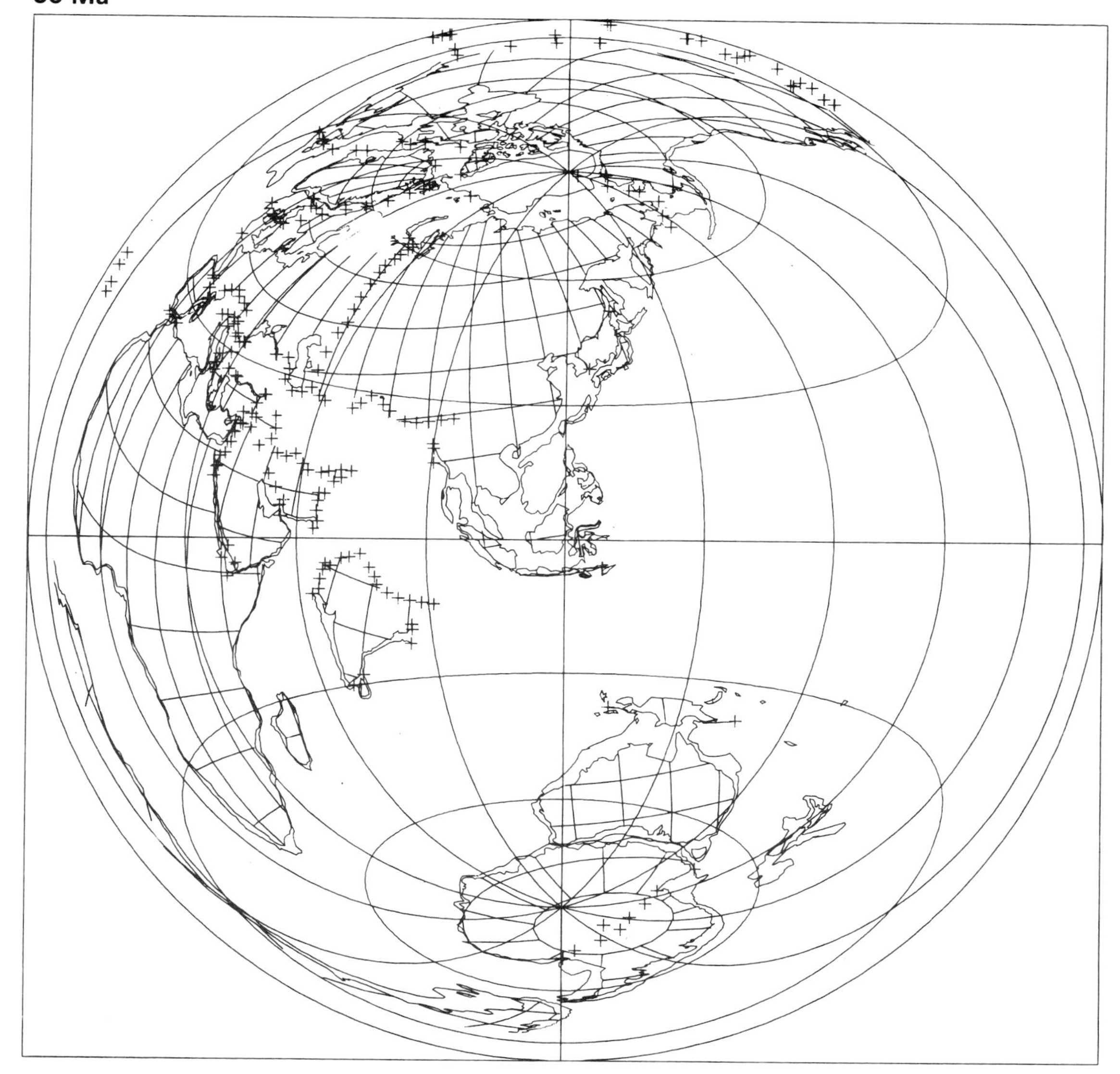

Fig. 3.7. The world at 60 Ma. The same movements continue. Australia/New Guinea will break away from Antarctica at about 55 Ma.

40 Ma

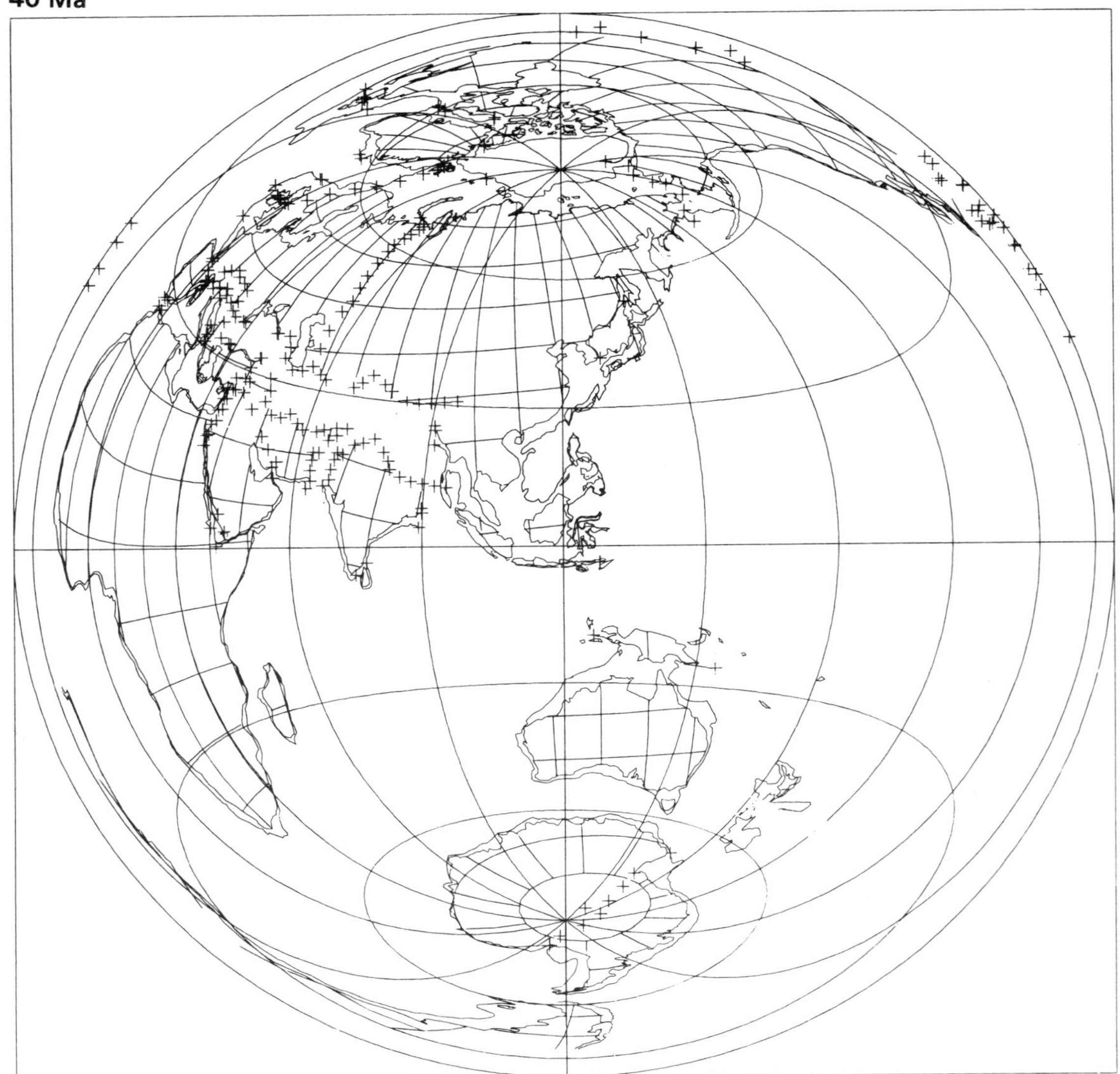

Fig. 3.8. The world at 40 Ma. India has collided with Asia (see text) and Australia/New Guinea has now separated from Antarctica and is moving northwards.

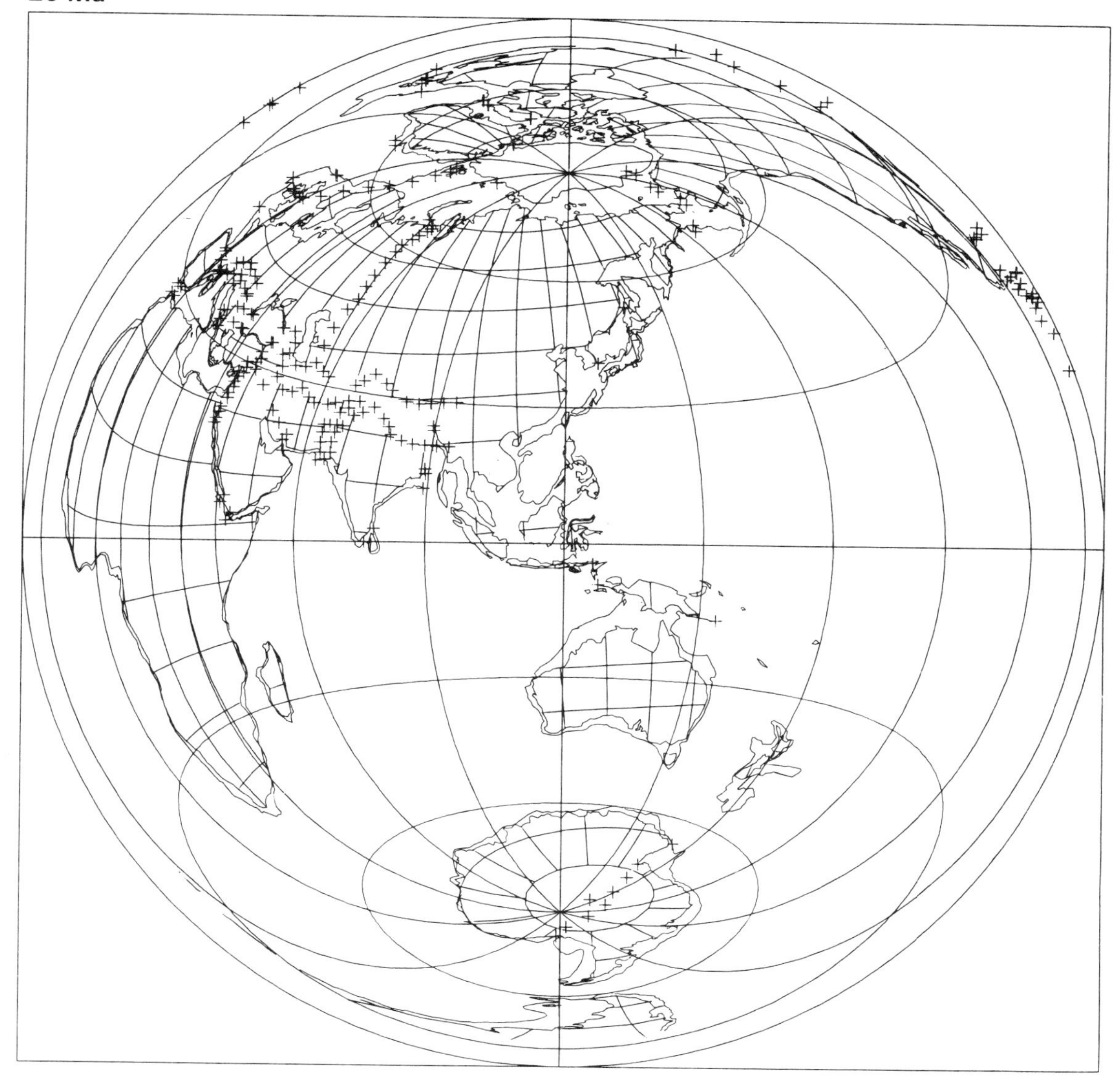

Fig. 3.9. The world at 20 Ma. Australia/New Guinea is approaching close to Asia but has not yet collided.

Today

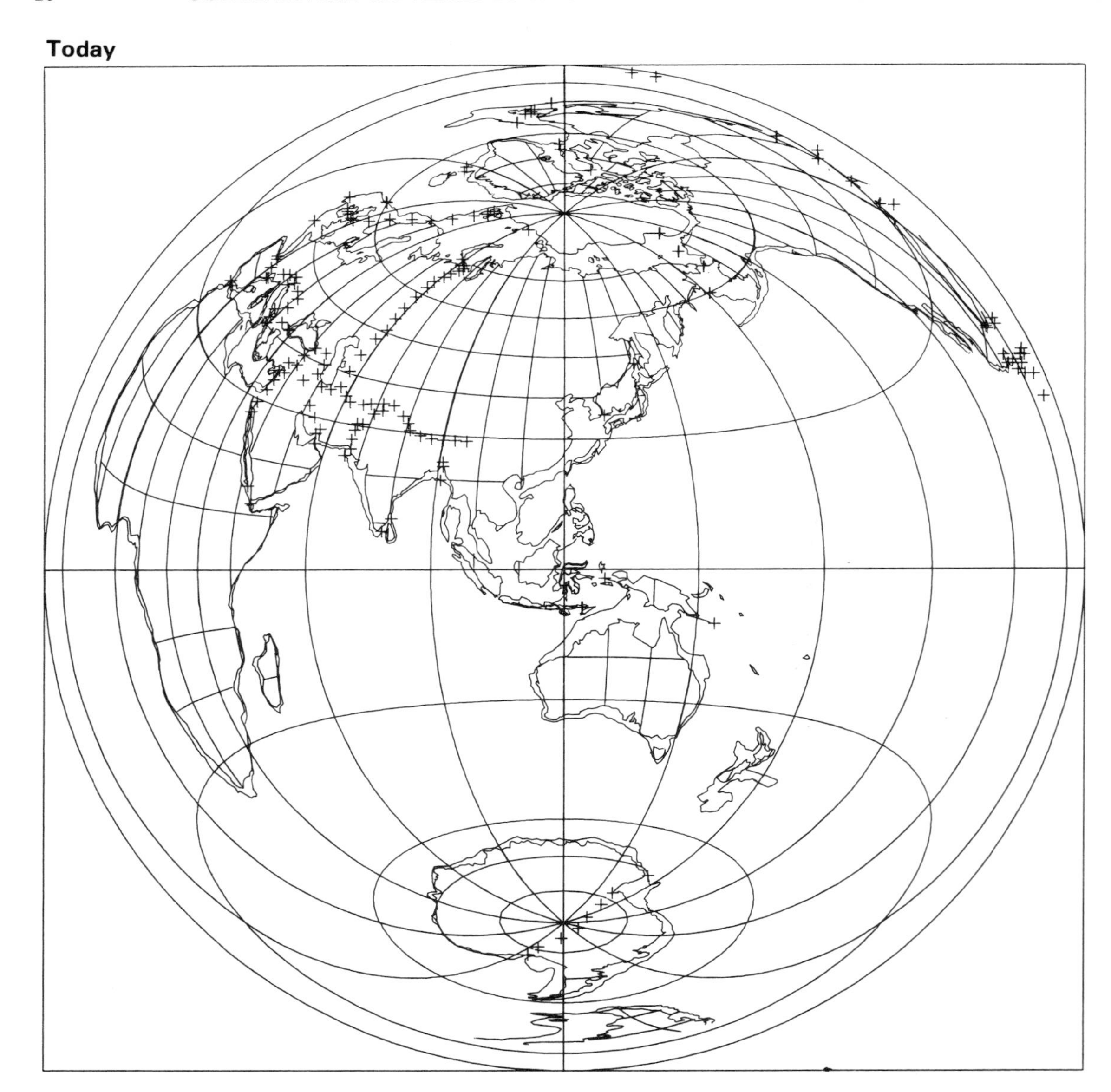

Fig. 3.10. The world today. The collision at Wallace's line is complete.

it was partially emerged than it is now, these islands were separated by at least 1000 km from them, and never formed a continuous land bridge. The effects, if any, of these islands on plant and animal distributions are minor and have not yet been detected.

Continental rafts

There is clear geological evidence that the continental margin of north-west Australia was formed by rifting in Triassic time. The piece that was rifted away has not yet been identified. The matter is discussed more fully below. Other possible rafts could also have existed on the northern edge of Gondwanaland and could have moved floras and faunas to new areas. None is likely to have been as large as New Zealand or New Guinea but, if they existed, their effects should be detectable in the biological record. In fact, geologists might have to use the biological record as evidence for such rafts, rather than geology itself.

Unscrambling deformed areas

Geologists can unscramble deformed areas, restoring them to their approximate original shapes. The global maps (Figs 3.1 to 3.10) have been corrected only in the region of Wallace's line (Figs 4.1 to 4.5). These qualitative modifications improve all maps prior to the present-day.

IMPORTANT FEATURES OF THE FRAGMENTATION AND NORTHWARDS DRIFT OF GONDWANALAND

Five features of the fragmentation and break up of Gondwanaland are important for the biogeographic evolution of the Malay Archipelago (Table 3.1). These can be best seen by examining this Table in conjunction with the maps (Figs 3.1 to 3.10).

Although the palaeomagnetic evidence suggests that Gondwanaland was in relative motion with respect to the northern continents throughout much of its history (e.g. Smith and Briden 1977), the supercontinent itself remained unbroken until about 140 Ma.

1. The first important event was at about 140 Ma when India probably separated from Antarctica–Australia, and from Africa and South America (see changes from Figs. 3.4 to 3.5). The precise dates are uncertain (Kidd and Davies, 1978; Norton and Sclater, 1979), but the separation of India from Africa must have begun prior to early Cretaceous time (Luyendyk, 1974). Although Veevers and McElhinny (1976) and Norton and Sclater (1979 speculate that India separated from Antarctica–Australia/New Guinea in early Cretaceous time we have separated them at the same time as India separated from Africa. Australia here includes much of New Guinea, part of eastern Indonesia and some of the southwest Pacific islands. All these areas will be referred to jointly as Australia/New Guinea.

India then began its northward drift, which at times was more rapid than any other continental movement so far known, 35–175 mm per year (Johnson, Powell, and Veevers 1976) (see Figs 3.5 to 3.8).

2. The second major event was the separation of Australia/New Guinea from Antarctica about 53 Ma ago (Johnson *et al.* 1976). A new ocean ridge formed between Australia and Antarctica which is still spreading at the present-day. The split between the two continents

Table 3.1

Principal continental movements important to the biogeography of the Malay archipelago

Figure	Million years	Era	Period	Epoch	
3.10	0		Q	Pleistocene	
		CENOZOIC	TERTIARY	Pliocene	④
3.9	20			Miocene	
				Oligocene	
3.8	40			Eocene	③
3.7	60			Palaeocene	②
3.6	80	MESOZOIC	CRETACEOUS	Upper Cretaceous	
3.5	100				
3.4	120			Lower Cretaceous	
3.3	140			Jurassic	①

Q = Quaternary

1. India probably splits from Antarctica – Australia/New Guinea and from Africa
2. Australia/New Guinea splits from Antarctica
3. India collides with Asia
4. Collision at Wallace's line of Asia and Australia/New Guinea (see Table 4.1)

took place between the times shown by Figs 3.7 and 3.8.

Although Gondwanaland did not start to break up until about 140 Ma, it had separated from the northern supercontinent of Laurasia (North America, Greenland, and Eurasia) at about 180 Ma by the opening of the central Atlantic Ocean between Africa and North America (Pitman and Talwani 1972). Laurasia itself broke up by the separation of North America from Greenland and of both from Eurasia (e.g. Smith and Briden 1977). The net effect of the break up and dispersal of both supercontinents was to close the Tethys Ocean that had originally lain between Gondwanaland and Laurasia in Triassic time. This closure resulted in a series of collisions between the northern edges of Africa, Arabia, Iran, and India and the southern edges of Eurasia.

Although not shown on the maps, Iran probably separated from Arabia in Permo-Triassic time, colliding with the southern edge of Eurasia in later Mesozoic time. It could therefore have been isolated from the major continents for a significant but not yet known interval of time, certainly sufficient to have brought about faunal and floral changes on it prior to its collision with Eurasia.

West of Iran there were probably numerous collisions between slivers of what are now Turkey, Greece, and Yugoslavia with the southern edge of Eurasia. The precise ages of these collisions are not well known and may span the Cretaceous and Tertiary periods. Most of these slivers may have been able to maintain connections between African and Eurasian marine faunas and floras, but the fossil record provides the best evidence for connections between terrestrial faunas and floras. Because Africa and Eurasia were close to each other throughout Mesozoic and Cenozoic time, it is likely that there was at least one migration route across shallow water open from Africa to Eurasia throughout most of Mesozoic and Cenozoic time, but the positions of such a route (or routes) is not yet well known.

3. East of Iran, an important event took place at about 55 Ma when India collided with Eurasia in middle Eocene time (Klootwijk and Pierce 1979) (Fig. 3.8). Unlike the collisions to the west of Iran, this collision had important implications for the present-day biogeography of the Malay Archipelago.

The north-west continental margin of Australia was block faulted in late Jurassic to early Cretaceous time—about 150–130 Ma ago—to form a new continental margin (Powell 1976). The ocean-floor spreading in the adjacent Wharton Basin is about 160–136 Ma old, roughly contemporaneous with the block faulting as one would expect (Falvey 1972, 1974; Veevers and Heirtzler 1974). However, the continental block that is believed to have separated from the Australian continent has not yet been identified. Veevers and Heirtzler (1974) suggested that the spreading ridge was 'initiated at or close to the previous continental margin along the edge of Tethys', presumably because the piece that broke off has not yet been located.

The ocean-floor anomalies in the Indian Ocean near north-west Sumatra and southern Java show that any collision of this presumed sliver with these areas must have taken place earlier than 75 Ma, i.e. latest Cretaceous time. It could perhaps have been subducted at the Java Trench or its ancestor.

An alternative possibility is presented here. In late Jurassic and early Cretaceous time, this presumed continental fragment of north-west Australia would have been adjacent to India (Fig 3.2). Thus the most logical place to look for a sizeable continental fragment that broke off from north-west Australia is to the north and east of present-day India. The geology of these areas does not appear to be sufficiently well known to evaluate this possibility.[1]

4. The convergence of Australia/New Guinea and Asia resulted in a collision of a different kind in the region of Wallace's line. This is an extremely young collision, taking place between 15 and 5 Ma—middle Miocene to Pliocene time—entirely between the times on the last two maps (Figs 3.9 and 3.10). This collision brought two originally separate faunas and floras into direct contact, ultimately giving rise to the present-day distribution of plants and animals that are the main topic of this book. The detailed geological effects are discussed in Chapter 4. In summary, the sinuous patterns of the Moluccan islands reflect the continuous collision during a ten-million-year interval of the irregular northern edge of the Australia/New Guinea continent with the island arcs to the north-west.

What this brief review of continental movements suggests is that there was no land connection between Australia/New Guinea and Eurasia from later Cretaceous time until the recent one that created the Malay archipelago. Evidence of possible earlier connections may be provided by the ranges of some of the flowering plants (Chapter 8).

[1] Mitchell (1981) identifies south Tibet and Burma–Thailand as former parts of Gondwanaland that were rifted away from that continent in late Permian or early Mesozoic (*c.* 200 Ma). However, the evidence for dating their separation from Gondwanaland is sufficiently uncertain to allow them to be considered tentatively as the missing continental fragment from north-west Australia.

HOW COLLISIONS ARE RECOGNIZED AND DATED

Collisions can occur between two continents, two island arcs, or a continent and an island arc. Ancient collisions may be identified by the geometry of reconstructions based on the ocean-floor magnetic anomalies. But the age of any collision is best given by the geological record in the collision zone itself. At least six kinds of evidence may be used to date collisions.

1. Prior to collision the edges of continental margins and island arcs may accumulate layers of undeformed sediment. The collision must be younger than the age of the youngest rocks belonging to these pre-collision sequences.

2. The deformation of rocks in the collision zone often takes place in a series of stages. The oldest rocks deposited after the oldest phase of deformation in a collision zone must be younger than the onset of collision.

3. Collision zones are usually marked by intense deformation. Rocks may be sliced up into large sheets that are stacked on top of one another; they may also be recrystallized to such an extent that the original textures and any fossils originally in the rocks have been obliterated.

4. Continents and island arcs generally move towards each other because the intervening ocean-floor is being 'subducted'—literally led underneath. The old ocean-floor, originally formed at a spreading ridge, is being recycled back into the mantle. During subduction, processes in the upper mantle partly melt rocks in the subduction zone. A portion of this molten rock may reach the surface as andesite (named after the Andes in South America). At the present time, andesites are a hallmark of subduction zones. When subduction ceases, andesites may continue to form for some time afterwards. Thus the age of youngest andesites is contemporaneous with or slightly younger than the age when the collision ceased. For example, volcanism in the islands of Alor and Wetar in the Inner Banda arc ceased in early Pliocene time because the Australian continental margin, now represented partly by Timor, had collided with this part of the arc (Audley-Charles 1980; Abbott and Chamalaun 1978; Figs 4.8 and 4.9).

5. Land bridges for the migration of terrestrial animals form after collision has begun. Thus the presence of placental mammals in India at about 55 Ma suggests that Asia, a possible source of such mammals, had collided with India at or somewhat earlier than this time (Colbert 1973).

6. Sediments are often deposited on one side of a collision zone from rocks that are being eroded on the opposite side. For example, northern India has thick sequences of granite detritus of middle Eocene or younger age whose source appears to have lain on the Asian side of the suture zone. Clearly, collision had taken place by middle Eocene time (Stoneley 1974).

As noted above, the present coastlines are shown on the world maps (Figs 3.1 to 3.10) even in areas like the Cenozoic volcanic island arcs that did not exist in Mesozoic time, and whose positions have changed throughout much of Cenozoic time. A second set of maps in Chapter 4 takes account of the geology of these and other areas whose shape or positions differed in past time. We now therefore turn to Chapter 4 for a more detailed reconstruction of the geological history of the region of Wallace's line itself.

4 GEOLOGICAL HISTORY OF THE REGION OF WALLACE'S LINE

M. G. Audley-Charles

There are now abundant indications from both geophysics and geology for lateral and vertical crustal movements in the region of Wallace's line. The first date of collision between Australia/New Guinea and the island areas off the Asian continent was mid-Miocene, about 15 Ma. The collision lay within the present island of Celebes, or just to its east. Extensive land was exposed above the sea between Australia and Celebes by the latest Miocene or early Pliocene providing a corridor for animal and plant migrations. The Makassar Strait between Celebes and Borneo has probably been a seaway since the Eocene at least but migration would have been possible when it was bridged, possibly in the mid-Miocene and very probably during the late Pliocene and Quaternary. The palaeogeography of the Philippines remains uncertain but part at least originates from the Asian continental margin and could have formed islands from 140 Ma (late Jurassic).

The region of Wallace's line is one of the most geologically complex zones known in the whole of the world. The present-day configuration of islands, island arcs, and basins filled with ocean waters to depths of between 3 km and 7 km is the result of complex crustal movements involving three different major events.

The key geological factor in determining possible migration routes for land animals and plants in the general region of Wallace's line has undoubtedly been the drifting of Australia/New Guinea northwards from its late Cretaceous position (*c.* 100 Ma) in the low southern latitudes, where it was continuous with Antarctica as part of Gondwanaland. This was described and illustrated in general terms in Chapter 3. Australia/New Guinea converged during the Cenozoic on the islands lying off the south-east Asian continental margin. The earliest collision occurred in the middle Miocene (15 Ma, event 5 on Table 4.1) and it provided an extended line of islands, joining Australia through central New Guinea to Celebes and thence westwards to mainland Asia, along which land animals and plants could have migrated. Figs 4.1 to 4.7 show details of the convergence and collision as reconstructed from all the evidence published up to mid-1979. Despite his disagreement with the structural interpretation of Carter, Audley-Charles, and Barber (1976), Hamilton (1979) follows closely the collision model for

Table 4.1

Principal geological events important to Wallace's line

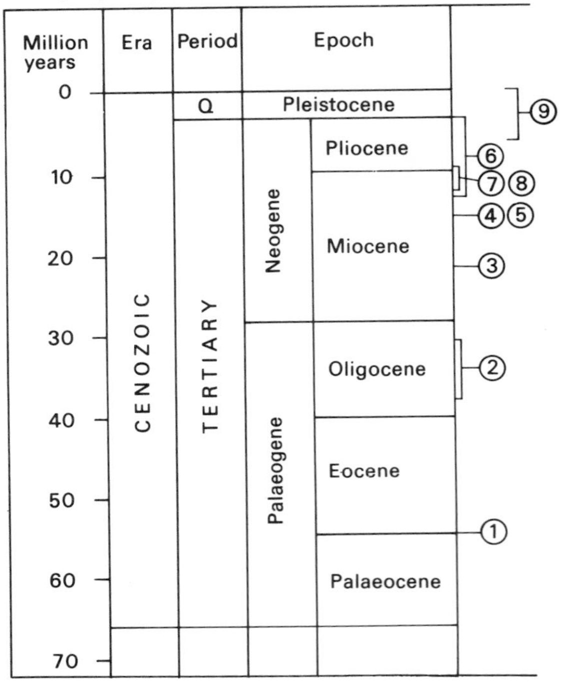

Million years	Era	Period		Epoch	Events
0	CENOZOIC	Q		Pleistocene	9
		TERTIARY	Neogene	Pliocene	6
10					7 8
				Miocene	4 5
20					3
30			Palaeogene	Oligocene	2
40				Eocene	
50					1
60				Palaeocene	
70					

Q = Quaternary

1. Australia/New Guinea splits from Antarctica (*c.*53 Ma).
2. Postulated formation of Philippines by collision of an Asian continental fragment with an island arc (Oligocene).
3. Possible land connection(s) across Makassar Strait (mid-Miocene).
4. Collision between New Guinea and a Tertiary island arc (*c.*15 Ma).
5. Collision between Gondwanaland (Sula Peninsula) and Laurasia at or near east Celebes (*c.* 15 Ma) but submarine.
6. Island chain established between east Celebes and Australia (late Miocene to late Pliocene).
7. Collision between parts of Gondwanic Outer Banda Arc and Laurasian (volcanic) Inner Banda Arc (latest Miocene to early Pliocene).
8. Gulf of Bone opens (latest Miocene to early Pliocene).
9. Probable land connection(s) across south Makassar Strait (from late Pliocene).

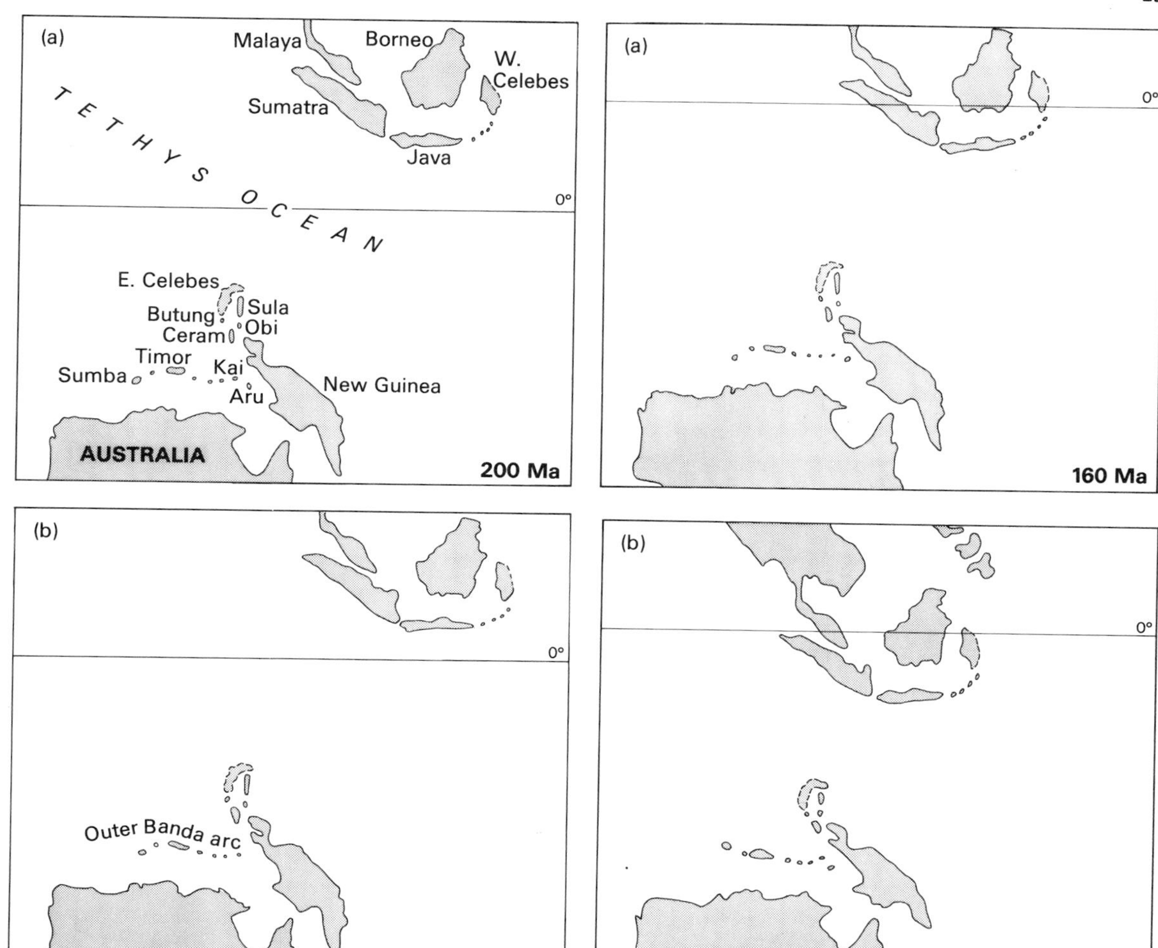

Fig. 4.1. The region of Wallace's line at (a) 200 Ma and (b) 180 Ma. Present-day coastlines are used for reference only as all palaeo-coastlines are very uncertain. This and Figs 4.2 to 4.5 are enlargements of part of the maps in Chapter 3 but updated to include geological information published to mid-1979. The equator is shown. Pecked lines indicate uncertainty over the palaeogeographical affinity of eastern Celebes and Butung.

Fig. 4.2. The same at 160 and 140 Ma.

the regions of the Banda Arc and Sula Spur-Celebes proposed by them and used in this chapter.

The second key event has been the movement of groups of islands away from continental Asia during the Cenozoic with the growth of small oceanic basins behind the migrating arcs (event 2 on Table 4.1; Dickinson 1973; Audley-Charles 1978).

Thirdly, there have been vertical crustal movements on faults, amounting to about 3 km displacement since the late Miocene (less than 15 Ma). As a result, islands have emerged and shallow seas and possibly land too have subsided several kilometres below present sea level.

EVIDENCE FOR LARGE-SCALE CRUSTAL CONVERGENCE

Through the greater part of geological history from about early Triassic (200 Ma) to about mid Eocene time (50 Ma), and while important continental movements were taking place elsewhere, the two halves of the Malay archipelago remained far apart, Figs 3.1 to 3.7 and 4.1 to

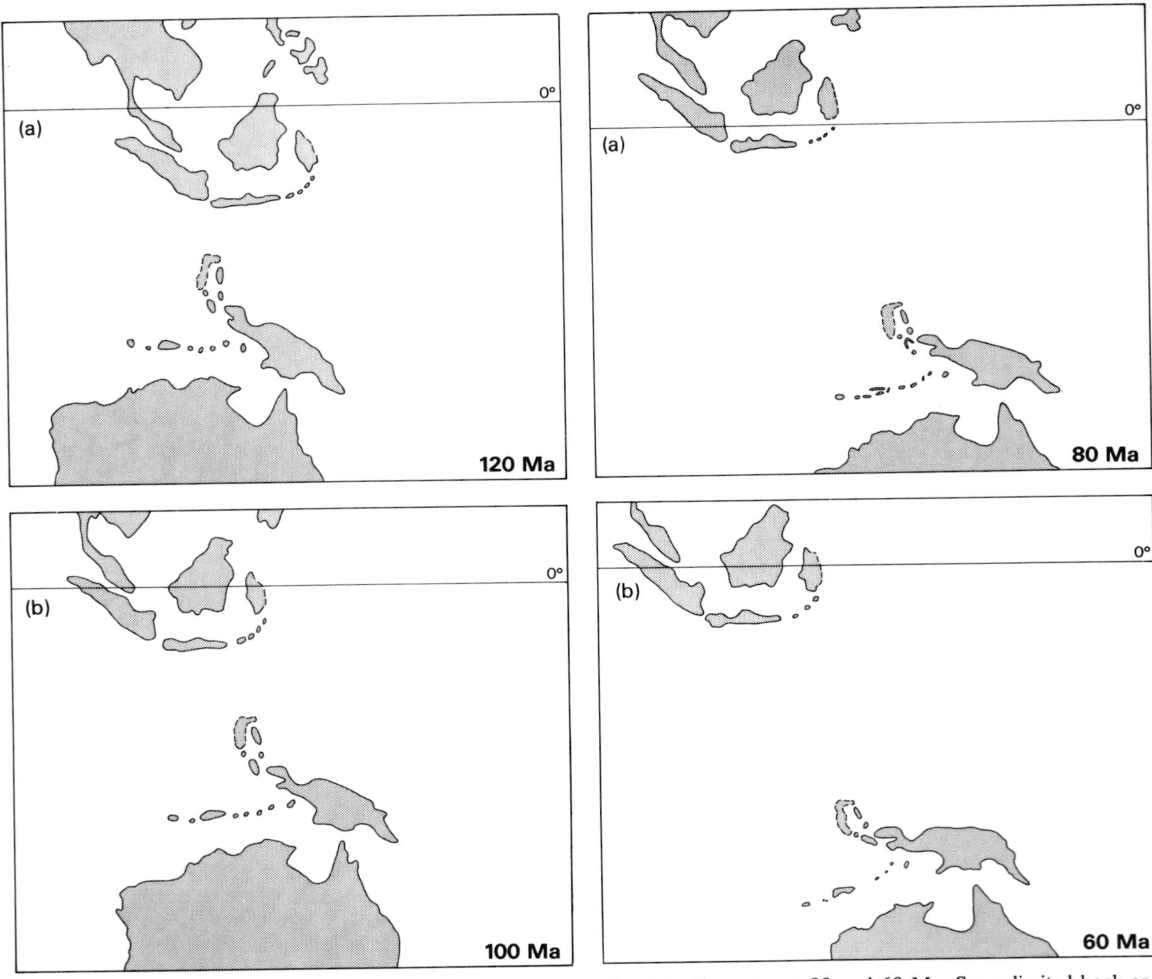

Fig. 4.3. The same at 120 and 100 Ma.

Fig. 4.4. The same at 80 and 60 Ma. Some limited back-arc spreading of the Banda Sea behind the proto-Banda volcanic arc may have occurred at this time but is not shown here (see Carter *et al.* (1976)).

4.4. Then, from 53 Ma, Australia/New Guinea moved northwards into the Tethys Ocean, rapidly converging on south-east Asia (Figs 3.8 to 3.10 and in more detail Figs 4.5 and 4.6).

There are four separate lines of evidence for this convergence.

The microfossils in the basal sediments covering the palaeomagnetically characteristic zones of the ocean floor indicate the growth and spreading of the Indian-Antarctic Ridge from 53 Ma (Johnston *et al.* 1976).

Secondly, palaeomagnetic data indicate a convergence between Asia and Australia during the Cenozoic, and there is similar evidence for spreading of new ocean south of the northward-drifting Australia/New Guinea (Smith and Briden 1977).

Thirdly, the present pattern of earthquakes between Australia and Indonesia (Hamilton 1974) indicate that they form a northward inclined zone. The hypocentres of the earthquakes deepen from less than 100 km in the region of the southernmost line of islands of eastern Indonesia (the Outer Banda Arc) to over 500 km depth below the Flores and south Banda Seas, to their north

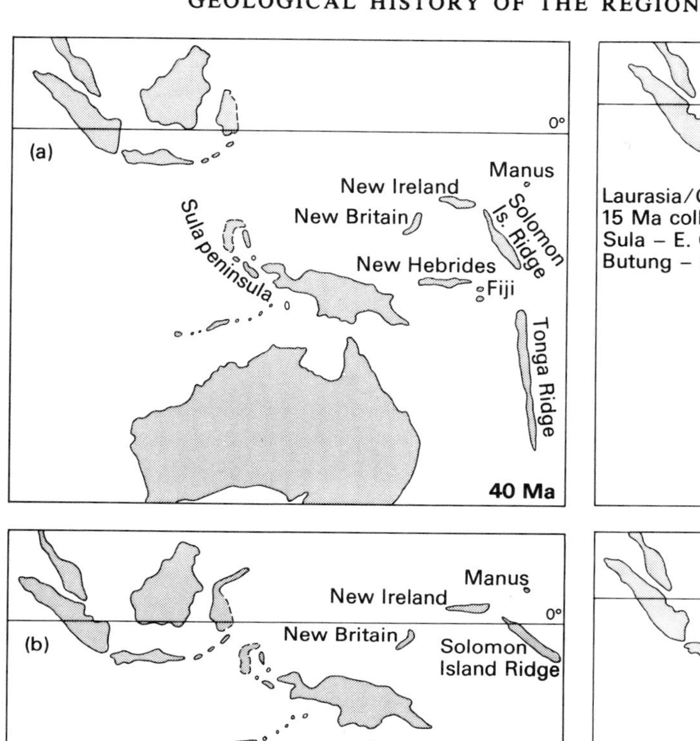

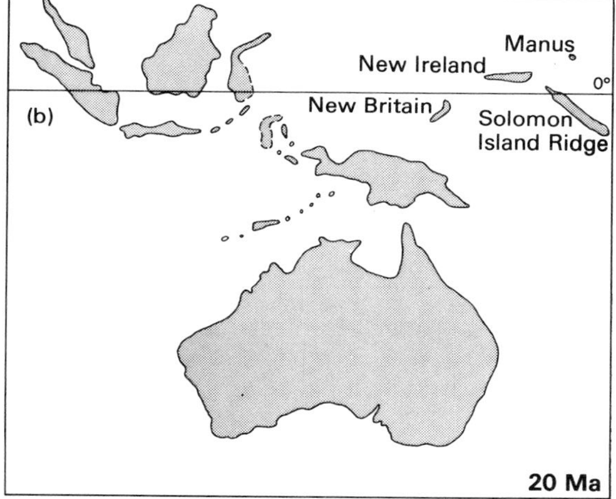

Fig. 4.5. The region of Wallace's line at 40 and 20 Ma. Convergence between Laurasia and Gondwanaland is now marked. Relative northward and westward movements of the Sula Peninsula (see Fig. 4.11) seem to be indicated and may be associated with the movements of the four plates: Asian, Australia/New Guinea, Pacific and Philippine Sea that meet in its region. The evolving Melanesian arcs can now be identified (Crook and Belbin 1978); their present-day coastlines are used for reference only.

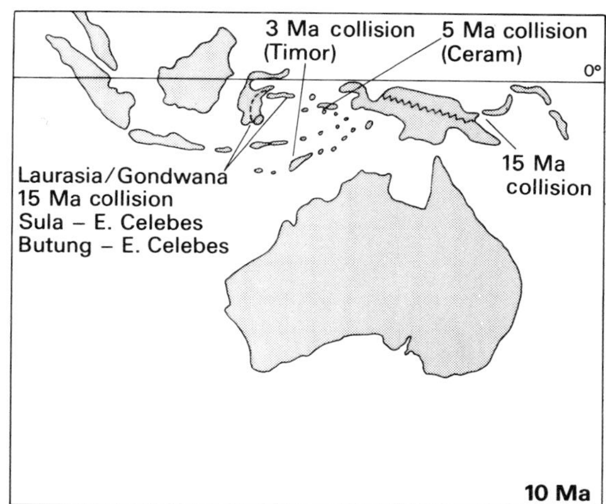

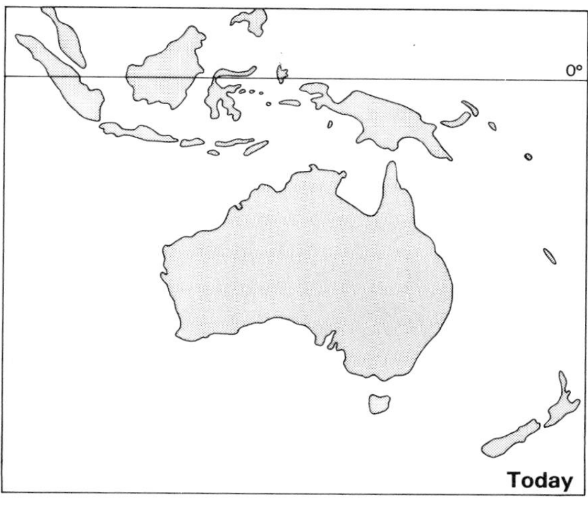

Fig. 4.6. The region of Wallace's line at 10 Ma and today. Present-day coastlines are used in the 10 Ma map for reference only. The collision of Sula Peninsula with Celebes is dated at 15 Ma. Continuing convergence led to collison between Ceram and the volcanic Inner Banda Arc in the late Miocene (about 5 Ma) and Timor collided with it in the early Pliocene (about 3 Ma). The convergence between Laurasia and Gondwanaland during the last 10 Ma in this region appears to have involved some rotation of the Vogelkop region of New Guinea and considerable strike-slip faulting in the collision zone of the Sula Peninsula. Both Norvick (1979) and Hamilton (1979), Fig. 77) follow the Banda Arc and Celebes–Sula Peninsula collision model of Carter *et al.* (1976) very closely. The New Guinea collision zone is detailed on Fig. 4.10.

(Fig. 4.8). A line of volcanic islands, the Inner Banda Arc, in which the oldest known rocks are early Miocene (*c.* 20 Ma; van Bemmelen 1949) are found about 150 km above this inclined zone of earthquakes. This indicates subduction of sub-ocean lithosphere belonging to the Australia/New Guinea plate below the Banda Island Arcs, and the cessation of volcanism in the early Pliocene

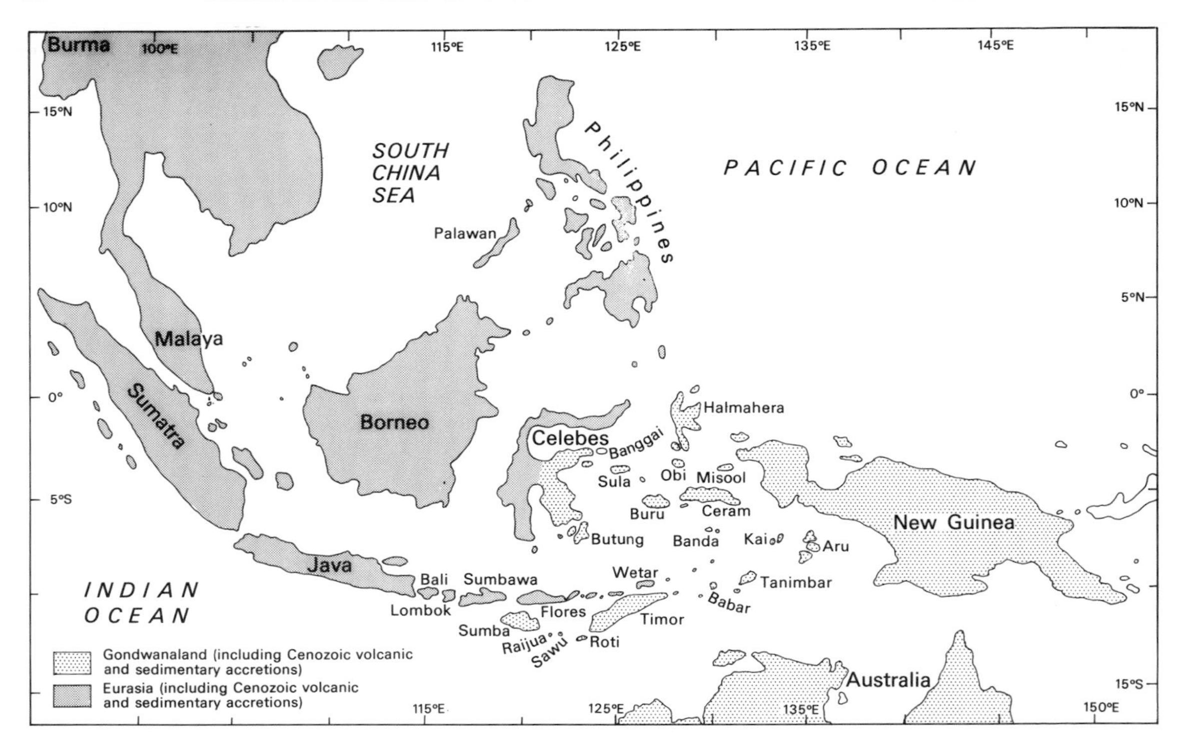

Fig. 4.7. The Malay archipelago today showing the main Gondwanic and Laurasian elements. In the Banda Arc where Laurasian elements overthrust the Gondwanic rocks only the latter are shown. Note that the Andaman and Nicobar islands, which are not included in this map, are thought ot have migrated oceanward by rifting from the Asian margin during the Neogene (last 11 Ma) with the spreading of the Andaman Sea back-arc basin (Curray *et al.* 1979). Thus their living flora is related to their geological history.

opposite Timor suggests collision then of Timor with Alor and Wetar, after all the oceanic lithosphere had been consumed by subduction.

Collision

The fourth line of evidence for Australia/New Guinea having converged on Asia during the Cenozoic is provided by the history of deformation in the Banda Arc at the margin of the Australia/New Guinea continent.

Locally, along this convergent zone between Australia/New Guinea and the island arcs of Indonesia, all the ocean crust that once lay along the northern margin of the Australia/New Guinea continent when it was part of Gondwanaland has been entirely underthrust back into the asthenosphere (Bowin, Purdy, Shor, Johnston, Lawver, Hartono, and Jezek 1980). The collision which then occurred, between the still northward-drifting Australia/New Guinea continent and the Inner Banda volcanic arc has resulted in deformation of the rocks in the collision zone. This deformation can be dated, thereby dating the collision. Collision zones tend to be uplifted, probably isostatically because the thickened low density sedimentary rocks tend to rise buoyantly above the denser rocks of the basement. Such vertical movements are accompanied by steeply inclined faulting, involving some blocks rising and others sinking.

In summary then, new islands emerge in collision zones by two processes. Volcanic island arcs build up before the collision. And non-volcanic islands arise in the front part of the collision zone where low density sedimentary rocks of great thickness and mechanically depressed into the denser crustal basement arc are uplifted by their buoyancy. These processes have in our case resulted respectively in the Inner and Outer Banda Arcs (Audley-Charles and Milsom 1974; Carter *et al.* 1976; Bowin *et al.* 1980).

Convincing geological and geophysical evidence has recently been published for the presence of Australian continental margin rocks below the Outer Banda Arc islands of Sawu, Roti, Timor, Leti, Babar, Tanimbar,

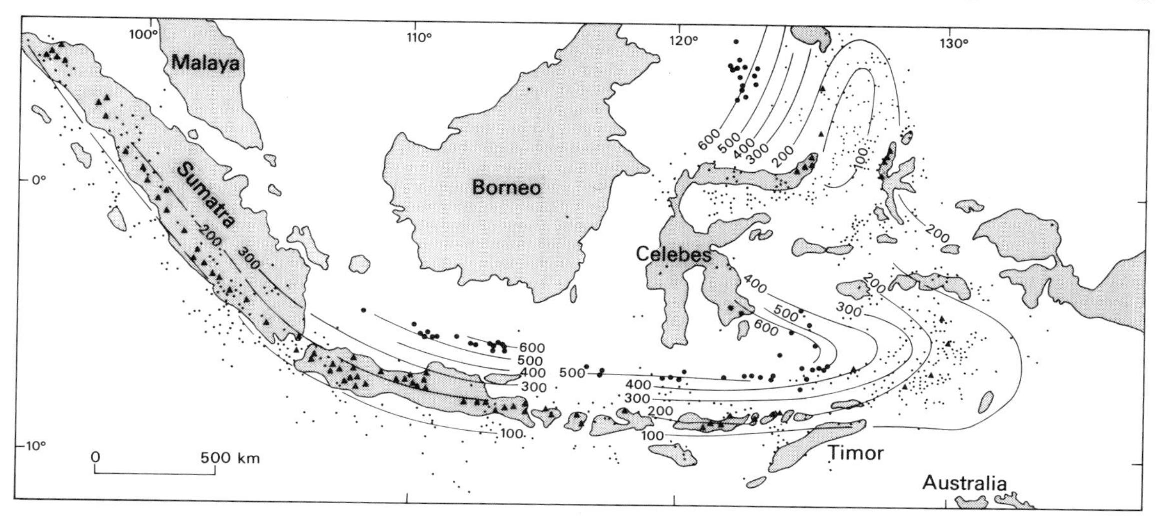

Fig. 4.8. Earthquakes and volcanoes in Indonesia (after Hatherton and Dickinson 1969). Circles: epicentres of earthquakes, large circles those of focal depth over 500 km. Triangles: volcanoes. Isobaths in km, on approximate centre of dipping seismic zone.

Kai, and the small islands between Kai and Ceram, Ceram itself, and Buru (Figs 4.7 and 4.9). This evidence comprises detailed similarity of rocks and fossils found in these islands and on the present Australian continental shelf and mainland (Audley-Charles 1978). Study of the earth's gravity field (which indicates the density of the crustal rocks (Chamalaun, Lockwood, and White 1976) and seismic velocity surveys have allowed these Australian strata and associated crystalline basement to be traced from the Australian shelf northwards below the Timor–Tanimbar–Ceram Troughs adjacent to the islands of the Outer Banda Arc listed above (Bowin *et al.* 1980).

By using the same kinds of evidence, it has long been established that central and southern New Guinea is underlain by Australian continental crust (e.g. by Sander and Humphrey 1975). Northern New Guinea appears to be underlain by continental margin 'transitional' crust locally overthrust by slices of oceanic crust called ophiolites (Bain 1973), resulting from collision at about 15 Ma of Australia/New Guinea with a Tertiary island arc that now forms the north coast ranges (Fig. 4.10) (Johnson and Jaques 1980).

Celebes

The key island in the biological discussion of Wallace's line is Celebes. It is also the critical island geologically for three main reasons. Firstly, western Celebes, now separated from Borneo by the Makassar Strait (locally 2 km deep and up to 250 km wide), may have been linked to Borneo by land, probably across the southern and central parts of the Strait, at various times during the Cenozoic and particularly during part of the Quaternary.

Secondly, Celebes appears to include several very different geological provinces suggesting the island is now composed of once widely separated groups of rocks brought together in the late Cenozoic time by large scale crustal movements. There is some geological evidence to suggest that part of eastern Celebes was formerly part of Australian Gondwanaland (Audley-Charles 1978) and this is supported by recently published palaeomagnetic evidence (Haile 1978; Sasajima, Nishimura, Hirooka, Otofuji, van Leeuwen, and Hehuwat 1980). However, neither the geological nor geophysical evidence is sufficiently detailed to be entirely convincing, and Norvick (1979), for example, suggests that all of eastern Celebes is a part of the same province as western Celebes, although he does not discuss the geological evidence and appears not to have seen the recent palaeomagnetic data. The suggestion made by Audley-Charles, Carter, and Milsom (1972) that Butung island at the southern end of the south-east arm of Celebes was formerly part of Australian Gondwanaland and part of the Outer Banda Arc geological province has since been supported by Sukamto (1975*b*), Katili (1975, 1978), Hamilton (1979), and Bowin *et al.* (1980). This correlation is based mainly on similarity of lithological and fossil sequences in Butung and in the Australian facies of the Outer Banda Arc, so that it needs the support of independent geophysical evidence before it can be regarded as reasonably well established.

The third, and most important, geological feature of

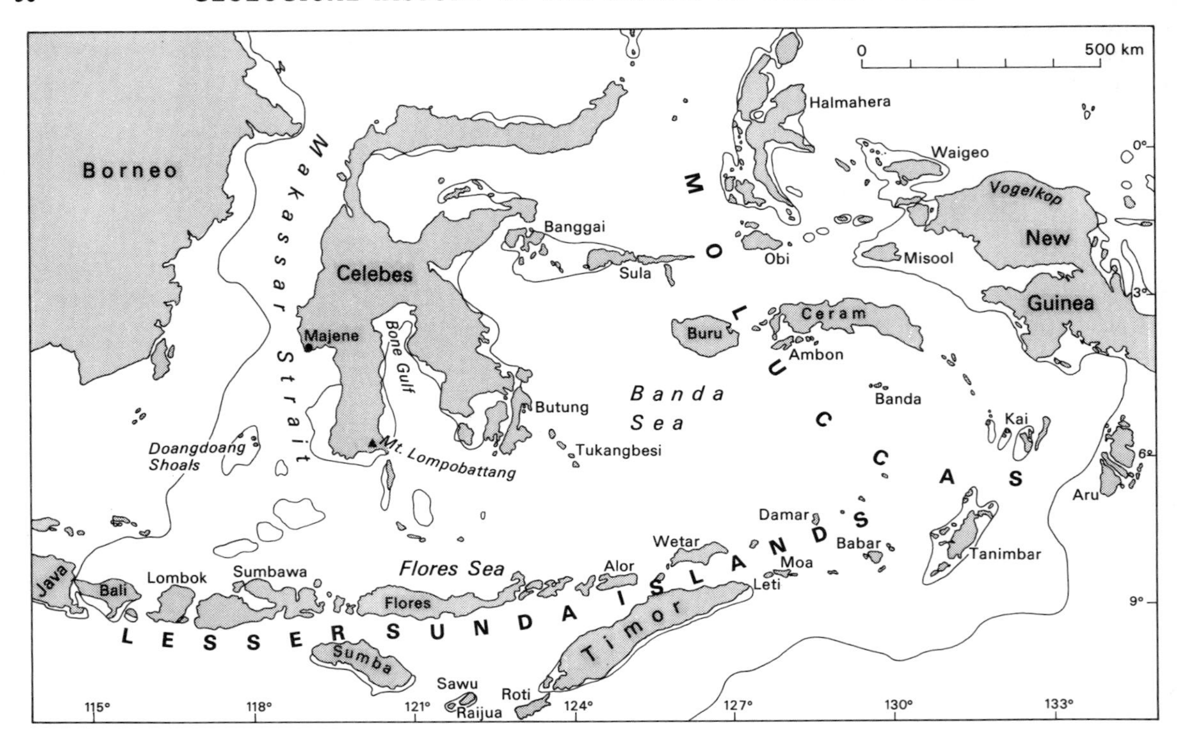

Fig. 4.9. The central part of the Malay archipelago to show the Lesser Sunda Islands, Moluccas, and (see text) the Banda Arcs. The 100 m (180 fathom) marine contour is shown, when sea level was lower the Makassar Strait would have been narrow at its south end.

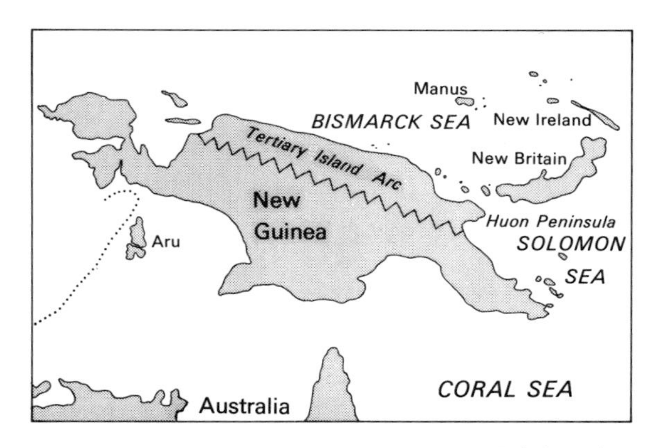

Fig. 4.10. Sketch map of the New Guinea–New Britain region. The 15 Ma collision zone and the north coast mountain range are indicated (Johnson and Jaques 1980). The approximate position of the collision suture is shown by a wavy line.

Celebes is the evidence for the collision between Celebes and what appears to be a detached slice of Australia/New Guinea forming the so called Sula Spur (Klompe 1956), reinterpreted as a much larger area by Carter *et al.* (1976) and called the 'Sula Peninsula' by Audley-Charles (1978) and Bowin *et al.* (1980) on the basis of additional geological and geophysical information (see Figs 4.9, 4.11). There appears to be very wide agreement among geological and geophysical investigators that during the late Cenozoic what now forms the Banggai islands (the westernmost exposed part of the Sula Peninsula) collided with what is now the east arm of Celebes (Klompe 1956; Audley-Charles *et al.* 1972; Hamilton 1979; Norvick 1979; Bowin *et al.* 1980). Although no argument appears to have been raised against this interpretation, it is based entirely upon geological sections. These have allowed correlations to be made between the rocks of the Sula Spur islands, which show they accumulated at the Australia/New Guinea continental margin, and the date of strong deformation of the rocks of the east arm of Celebes (Kundig 1956), Ceram and other islands of the Outer Banda Arc. The collision between the Sula Peninsula fragment of Australia/New Guinea and the east arm of Celebes has been dated as middle Miocene on the basis of the age of overthrusting and imbrication in the east arm of Celebes (Kundig 1956). This collision provided the first direct connection between the Australian continent and south-east Asia.

Was the collision zone above the sea?

For the purposes of discussing the potential migration route for land plants and animals the next geological question that needs to be answered is: did the collision between the Sula Peninsula and east Celebes involve a land or submarine connection? This enquiry must be addressed to both the Asian and the Australian side of the collision suture.

Geological evidence indicates that parts of east Celebes appear to have emerged from a submarine to a land environment immediately after the collision because, according to Kundig (1956), the deformed (imbricated) rocks of the collision zone range in age from late Cretaceous to middle Miocene marine sediments and they are overlain unconformably by middle Miocene to Pliocene molasse (erosion detritus). The presence of this molasse and also of Pleistocene terraces of raised fringing reefs testify to at least parts of eastern Celebes remaining above sea level after the post-collision emergence.

On the Australian (Gondwanan) side of the collision suture, there is some weak evidence from coral reefs and lignites that the Banggai and Sula islands on the western part of the Sula Peninsula (Figs 4.9, 4.11) have been above sea level since the Neogene. Following the middle Miocene collision, coral reefs of late Miocene to Quaternary age accumulated (Sukamto 1975*a*). This does not definitely indicate whether western parts of this Peninsula were above sea level then but only shallow marine waters are certainly implied. Furthermore, the geographical distribution of these reefs suggests, especially by analogy with the other islands of the Peninsula, that they developed around islands where older rocks are now exposed (Sukamto 1975*b*). There is other evidence too. Van Bemmelen (1949) described bedded lignites and conglomerates of late Miocene and Pliocene age from these islands. By analogy with the late Neogene lignites and associated conglomerates in the nearby island of Ceram and the geologically related island of Timor, these deposits suggest (but do not prove) that land was exposed while they were accumulating.

Land connection to Australia

We need to follow the Sula Peninsula south-eastwards

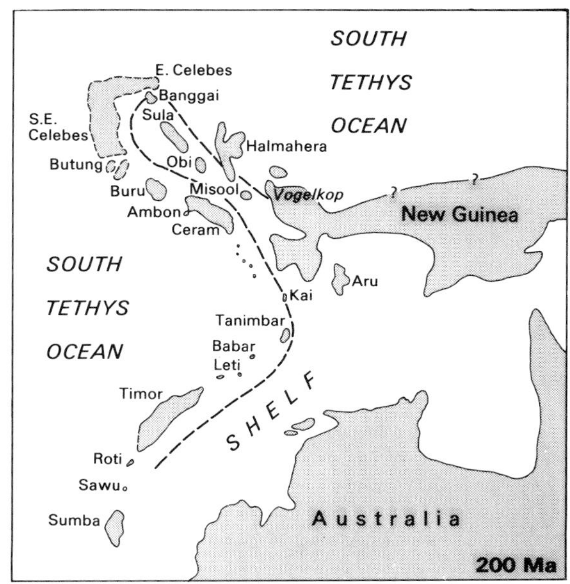

Fig. 4.11. Qualitative reconstruction of the Sula Peninsula projecting north-west from Australian Gondwanaland at 200 Ma (from Carter *et al.* 1976 and Audley-Charles 1978). Note this reconstruction involves some local rotation in the Vogelkop region of New Guinea and opening of the Banda Arc in the Buru–Ceram region, an interpretation subsequently followed by Hamilton (1979, Fig. 77). The uncertainty over the palaeogeographical affinity of east Celebes and Butung is indicated by pecked lines. The present-day line of islands from east Celebes, Buru, Ceram, Tanimbar, to Timor (i.e. present Outer Banda Arc) formed part of the Australian continental rise from early Permian until the late Cenozoic collision. The interpreted position of the pre-collision shelf break is indicated.

into New Guinea and eastern Australia via the present islands of Buru, Obi, Ceram, and Misool (Fig. 4.11) to investigate the evidence for land having existed there during the late Miocene and Pliocene, as this would have been the most direct potential migration trail for land plants and animals.

Recent findings suggest that Ceram was entirely submarine during the pre-late Miocene (Audley-Charles, Carter, Barber, Norvick, and Tjokrosapoetro 1979). The collision between the Sula Peninsula and the Inner Banda volcanic arc in the Ceram and Ambon region has been dated as latest Miocene or earliest Pliocene (biostratigraphical zones N.18–N.19). There is clear evidence that by late Pliocene times parts of the present Ceram island were already above sea level. It is possible that some of the present mountainous area of central Ceram had emerged by early Pliocene and perhaps even earlier in the late Miocene immediately after collision, when the great thrust sheets were emplaced in the western central spine of the present island.

Buru is much less well known geologically but van Bemmelen (1949) referred to Plio-Pleistocene conglomerates possibly of alluvial origin, which would indicate that parts were above sea level before the end of Pliocene and perhaps earlier. Obi and Misool seem, on the available rather sketchy information provided by fringing reefs and alluvium (van Bemmelen 1949; Froidevaux 1974), to have emerged only in the Quaternary, but the precise timing is uncertain and the possibility of some Pliocene land here cannot yet be ruled out.

In the maps (Figs 4.5, 4.6) the palaeogeographical distribution of land and sea has been indicated by plotting present day coastlines of the islands, whereas the area above sea level before 20 Ma must have been considerably less than at the present time. The precise position of these islands is uncertain but on the scale of these maps they are not likely to be substantially displaced. To locate them more accurately would require details of the spreading history of the intervening ocean floor of these marine basins. Bowin *et al.* (1980) suggest the floor of the Banda Sea may not be amenable to this kind of study because of its complex structure.

Land connection summarized (Table 4.1)
The possible consequences, in terms of providing potential migration routes for land plants and animals, of the collision between the Sula Peninsula and eastern Celebes may now be summarized (see Figs 4.9, 4.11).

1. The earliest collision (Fig. 4.6) between the Australia/New Guinea continent and an Asian land mass (Celebes) has been dated as middle Miocene. At that time, or almost immediately afterwards within the middle Miocene, there was land exposed in east Celebes.
2. Further south and east in New Guinea huge tracts of land two thousand kilometres long had emerged above the sea by middle Miocene times along the central mountainous spine (Thompson 1967; Froidevaux 1974).
3. At the time of this collision and immediately preceding it parts of the Sula Peninsula including Buru, Ceram, and Misool appear to have been submarine but Banggai, Sula, and Ceram possibly began to emerge by late Miocene or early Pliocene time and certainly by late Pliocene, and Buru possibly by late Pliocene also.
4. During the late Pliocene and Quaternary fluctuations of sea level, resulting from changes in the polar ice sheets, could have exposed extensive tracts of land at some of these islands such as Ceram and Buru. They were also, and still are, rising isostatically as a consequence of the mechanical deformation they suffered during the late Miocene–early Pliocene collision.

The suggestion strongly emerges that the earliest migration route for land plants (and possibly animals) could have been established between Celebes and eastern Australia by the latest Miocene or early Pliocene and that by the late Pliocene it was probably almost as well established as at present.

As mentioned above there is some evidence to suggest that part of east Celebes and Butung actually formed part of the Sula Peninsula of Australia/New Guinea (i.e. part of Gondwanaland) throughout the Mesozoic and collided with the Asian part of eastern Celebes in the middle Miocene. At present this remains somewhat speculative so is indicated on Figs 4.1 to 4.6 and 4.11 by pecked lines, although recently published palaeomagnetic evidence (Sasajima *et al.* 1980) supports this interpretation. The implications of this geological model need to be borne in mind in discussing details of plant and animal distributions within Celebes.

Late Pliocene and Quaternary sea levels would also have united Sumatra, Malaya, Java, and Borneo, all of which lie in the shallow seas of the Sunda continental shelf.

THE OPENING OF THE MAKASSAR STRAIT SEPARATING CELEBES FROM BORNEO

Parts of the Makassar Strait (Fig. 4.9) overlie at least 3·5 km thickness of Cenozoic sediment (Koesoemadinata and Pulunggono 1975). There are also reports of marine sedimentary rocks of Eocene and middle Miocene age (Hamilton 1974). These discoveries suggest the Makassar Strait has been a marine basin of sedimentation, locally at least, since the Eocene and, in view of the deep-water Cretaceous sediments of south-east Borneo and western Celebes, possibly extending locally back into the Cretaceous. Deltas have been discharging large bodies of sediment into the Strait from Borneo since part of Miocene time (Matharel, Klein, and Oki, 1976). Further, the geological map of the region indicates marine basins along the margins of south-east Borneo and western Celebes, suggesting that the Makassar Strait has been a seaway since the Eocene at least. Despite these observations the possibility of local islands emerging during the Cenozoic between Borneo and western Celebes cannot be ruled out, especially in the region near Majene and of the Doandoang shoals at the southern part of the Strait where at present water deeper than 180 m (100 fathoms) is confined to a narrow passage. Sea levels lower than present by less than 180 m would expose almost continuous land between south-east Borneo and south-west Celebes.

In fact the changes in sea level known to have occurred during the late Pliocene and Pleistocene would have been sufficient to provide such land. Furthermore, there are geological grounds for thinking that, during the middle Miocene orogenesis that affected central Celebes, land connections could have been well established across the Strait. There does not however appear to be any published geological information that allows this to be interpreted with a high level of confidence.

Katili (1978) argued that the Makassar Strait was open during the pre-Pliocene, closed during the late Pliocene, and then reopened by post-Pliocene rifting. Similarly Hamilton (1979) suggested that Celebes rifted from Borneo in the middle Palaeogene (about 45 Ma) thus opening the present Makassar Strait. These views do not find support in the published information and are at present no more than interesting speculations.

Another important question to be answered is from what time in the Cenozoic was land exposed in western Celebes? The answer seems to be provided by the presence of terrestrial volcanic rocks and alluvium that date back to the late Miocene (Sukamto 1975*a*), proving that some land was exposed in western Celebes at least that far back.

OPENING OF THE GULF OF BONE

The Gulf of Bone (Fig. 4.9) separates the two southern arms of Celebes. It thus occupies an analogous geological position to central Celebes which separates western (Asian) Celebes from eastern Celebes, which together with Butung, perhaps (see above) contains some Australian elements belonging to Gondwanaland. However, some geologists might not accept this last point and might follow Norvick (1979) in regarding eastern Celebes as entirely Asian. Even on this view central Celebes is still geologically a structural suture representing a zone of intense compressive deformation (Bowin *et al.* 1980). The problem highlighted by Norvick's interpretation is to explain the presence of a gulf filled with 2 km of seawater between the two southern arms of Celebes. One of two possibly related processes can be invoked. Either the Gulf of Bone is considered to be a rift zone where the originally contiguous two southern arms of Celebes have been pulled apart (Hamilton 1978; Norvick 1979) or alternatively it could have resulted from down-faulting between what are now the southern arms. Either model would require the formation to have been post-middle Miocene, after the compressive forces associated with the collision of the Sula Peninsula had deformed the rocks of central Celebes. A late Miocene–early Pliocene age is also suggested by the distribution of marine sedimentary basins around the margins of the gulf (Sukamto 1975*a*).

MOUNT LOMPOBATTANG

According to van Steenis (1972*a*) Mount Lompobattang (Bonthain) of south-west Celebes (Fig. 4.9) supports a flora that appears to be much more closely related to that of the Lesser Sunda Islands (which geologically includes part of the Outer Banda Arc) than to the floras of western Celebes, Java, or Borneo. One possible explanation is provided by the geology. The collision between the Sula Peninsula and eastern Celebes (Fig. 4.11 and above) could have provided an immigration route to Lompobattang for the Outer Banda Arc floras through islands such as Timor and Ceram which had certainly emerged by late Pliocene times. Or alternatively the floras of both Mount Lompobattang and Timor and the other Lesser Sunda Islands could have immigrated from New Guinea which contained an extensive land area by middle Miocene times. But why are such floras now found in Mount Lompobattang and not in other parts of south-west Celebes? This may perhaps be because this mountain comprises Miocene, Pliocene, and Quaternary volcanics (Sukamto 1975*a*) and may have been an isolated island in western Celebes when most of that territory was covered by the sea in the late Miocene and early Pliocene. The collision of the Sula Peninsula with eastern Celebes (which it must be remembered occurred before the formation of the Gulf of Bone) would have provided a fairly short and direct migration route for land plants to have reached Mount Lompobattang in the late Miocene–early Pliocene. The later emergence of other parts of western Celebes from below the sea at the same time that land flora was able to migrate there from Borneo, via the southern part of the Makassar Strait discussed above, could have isolated the Mount Lompobattang flora.

EVOLUTION OF THE LESSER SUNDA ISLANDS

The age of emergence as islands of the volcanic archipelago east of Java (Figs 4.8, 4.9) is uncertain. The oldest rocks have been dated as early Miocene (van Bemmelen 1949) on the basis of marine fossiliferous sedimentary rocks interbedded with the volcanics. It seems unlikely that older rocks will be discovered in this area of the Indian Ocean floor because the oldest rocks now reaching the Java Trench south of Java is late

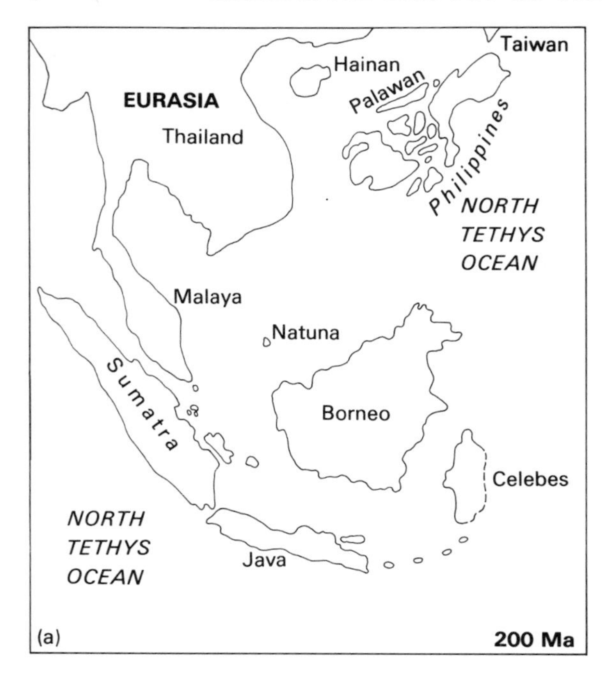

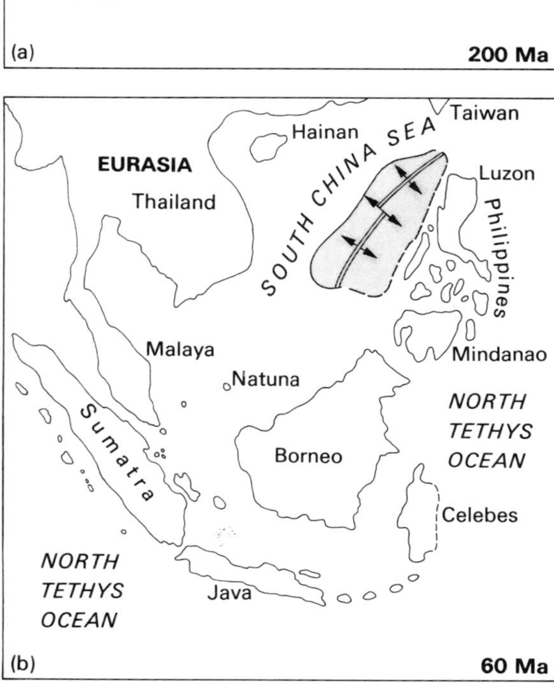

Fig. 4.12. (a) Palaeogeographical sketch map for 200 Ma to show one interpretation of the position of the Philippines (after Audley-Charles 1978). Present-day coastlines are for reference only. Pecked lines in Celebes indicate uncertainty of palaeogeographical affinity of its eastern region. (b) The same for 60 Ma. Note the postulated spreading in the South China Sea with oceanward migration of the Philippines (after Audley-Charles 1978). Position of the postulated intraoceanic arc involved in collision with parts of western Philippines in mid-Cenozoic (Roeder 1977) is not shown.

Cretaceous (Johnson *et al.* 1976), older floor having been subducted.

The size of the islands of this volcanic chain gradually diminishes eastwards from Java through Bali, Lombok, Sumbawa, Flores, Wetar to Banda (Fig. 4.9). This diminution, which is most noticeable east of Wetar, may reflect the amount of ocean floor subducted, implying either that dip-slip motions have been more important westwards from Wetar and strike-slip motions increasingly important eastwards. Alternatively, it may be that the present volcanic arc east of Wetar is younger and perhaps that the original volcanic arc east of Wetar has been overridden by the Australian continental margin (Bowin *et al.* 1980).

From the point of view of potential migration routes for land plants and animals it seems likely from what is known of Wetar (Abbott and Chamalaun 1978), that this part of the arc formed islands by the late Miocene and that the volcanic island chain provided a potential migration pathway of islands significantly smaller than they are today (great uplift occurred during the Plio-Pleistocene) to connect with Java by the end of the Miocene (van Bemmelen 1949). This means that the volcanic island chain could have been colonized in the early Pliocene by any land animals and plants migrating eastwards from Asia through east Java which were able to cross the strait between Bali and Lombok which was then probably narrower and shallower. As described above, the Australia/New Guinea continent collided with the eastern part of this volcanic island chain during the latest Miocene/early Pliocene so that migrations into the volcanic islands westwards from Gondwanaland may also have been possible from the early Pliocene (Table 4.1 event 7). By the middle Pleistocene animals such as pygmy stegodonts from Java had migrated from Laurasia on to the Gondwanaland margin in Timor by walking on dry land along the volcanic island arc of Flores–Wetar (Chapter 7; and Audley-Charles and Hooijer 1973). The volcanic islands of Flores, Alor, and Wetar are now separated from Timor by a strait 3 km deep which formed by down faulting in the late Pleistocene.

THE PHILIPPINES

The Cenozoic palaeogeography of the Philippines (Fig. 4.12) remains uncertain and can only be described in speculative terms. Two principal palaeogeographical models have been proposed:

1. The first view regards the Philippines as a fragment rifted away from the mainland of south-west China during the late Jurassic of early Cretaceous (Audley-Charles 1978) with the spreading of the South China Sea as a new marine basin (Ben Avraham and Uyeda 1973). There is some geological evidence for this model but not enough to put it forward with confidence. One difficulty is that most of the evidence of this postulated Mesozoic ocean floor spreading would have been destroyed by the subduction concomitant with the later Cenozoic spreading.
2. The second model regards the Philippines as the product of a collision between two intraoceanic island arcs (Roeder 1977). There is some geological evidence to support this too.
3. The author now favours a third model comprising elements of the first and second which regards the Philippines as having a composite origin. The western parts are regarded as fragments rifted during the late Mesozoic from the continental margin of south China while the eastern parts are considered to be an intraoceanic arc that collided with the rifted fragments during the Oligocene (Roeder 1977).

The question that arises in the case of any of these palaeogeographical models is when was land exposed in the Philippines that could have been colonized by land plants and animals? To this it is very difficult to give a confident answer. The best available evidence seems to suggest that some Philippines volcanoes can have been above sea level for about 140 Ma, since about late Jurassic times, when they were close to the Asian continent (Audley-Charles 1978), and other volcanoes and associated islands may have been above sea level for 70 Ma since the late Mesozoic.

WESTERN PACIFIC ISLANDS

The composite model for the origin of the Philippines favoured here provides migratory pathways between the rifted fragments of Asia, and hence northern Borneo, and island arcs of the western Pacific during post-Oligocene times.

Another migratory pathway with islands of the western Pacific, this one on the Gondwanaland side of the regional suture, was provided by the collision of the stable platform of New Guinea with a volcanic island arc during the late Oligocene to Miocene (*c.* 15 Ma) (Figs 4.6, 4.10 (and above)). This collision formed the northern coastal range of New Guinea (Johnson and Jaques 1980).

SUMMARY, Table 4.1

The outstanding new geological discovery that bears on the position and significance of Wallace's line is the detailed evidence for the first collision between the Sula Peninsula extension of Australia/New Guinea with east Celebes during the middle Miocene. This provided, possibly by late middle Miocene times or by the late Miocene and certainly by the late Pliocene, a line of islands (some very large) extending from east Celebes through Banggai and Sula, Ceram and New Guinea to eastern Australia (Figs 4.9, 4.11). Land plants and animals would have been able to migrate both ways along this emerging island chain between Celebes and Australia.

Western Celebes seems to have been separated from Borneo by a seaway forming the Makassar Strait since the Eocene or Cretaceous except perhaps at the southern end, where, via the Doandoang shoals region, a line of islands and/or even extensive land may have allowed land plants and animals to migrate from Borneo into south-west Celebes during the late Cenozoic (Fig. 4.9). Land seems to have been exposed in western Celebes since the late Miocene and possibly earlier.

The Lesser Sunda Islands could have been colonized from both the east and west by the early Pliocene across short sea distances and by middle Pleistocene there was dry land from Laurasian Java to Gondwanic Timor.

Other geological data suggest (but do not yet prove) that the Philippines could have provided a migration route for land plants and animals from south China into northern Borneo and Celebes since the middle Cenozoic and could possibly have also been a migration route for interchange of land plants and animals with intraoceanic island arcs of the western Pacific after about Oligocene time. Other connections with the western Pacific islands developed in the region of New Guinea during the late Neogene.

5 PALAEOCLIMATE AND VEGETATION HISTORY

T. C. Whitmore

Animals and plants are variously confined to particular habitats. There is now good evidence that the climate of the tropics has fluctuated continuously during the Tertiary and Quaternary and that this has altered the relative extents of the different sorts of forest, hence the ranges of individual animal and plant species. Present-day distribution patterns may reflect past climatic vicissitudes. In the tropics the climate during much of the Pleistocene has been drier and more seasonal than the present day and the sea level has been lower. It is likely that the two great rain forest blocks of Malesia, centred on Sundaland and Papuasia respectively, have been of somewhat reduced extent and seasonal forests expanded. This must be allowed for in any attempt to correlate species' ranges with palaeogeography. The chapter ends with a speculative look at the floristic interdigitation which took place in the late Tertiary after the Laurasian and Gondwanic floras came into contact.

Excellent analyses and reviews of Quaternary climate and vegetation for the tropical Far East have been published by Walker (1972, 1981), Verstappen (1975), and Flenley (1979). Here it suffices to give a summary. The evidence remains fragmentary and, although nearly all the data on vegetation are from the mountains, a general picture can be built up.

PALAEOCLIMATE

During the Tertiary it is likely that the decrease in temperature from the equator to the poles was considerably less than today. Also the poles were warmer. Tropical and subtropical climates extended further away from the equator and this arose in part from the different dispositions of the large land masses (Smith 1977; Chapter 3). Tropical and subtropical plants and animals would have been able to live further away from the equator than at present and this affected their ability to survive and migrate on the evolving and moving continents and islands. The warm Tertiary climate, therefore, influenced the fauna and flora which reached what is today the Malay archipelago, and was important in the determination of the major patterns of evolution.

But the most important climatic changes which have modulated the present day ranges of plant and animal species are those of the Quaternary, occupying the last two million years. Detailed information has been obtained for the late Quaternary and is perhaps indicative of the changes which took place continuously through the whole period. There is now good evidence that the late Quaternary Glacial periods of high latitudes were times of drier climate (Manabe and Hahn 1977) and lower sea level and that the Arctic and Antarctic glaciations were in phase. The cooling of what are temperate areas today locked up a huge quantity of water as ice. Sea level was eustatically lowered by as much as 180 m below the contemporary level in the Malay archipelago. At 11 170 BP rainfall may have been as much as 30 per cent below the present day amount in the equatorial zone though globally it averaged only about 20 per cent lower. The global cooling affected tropical mountains and depressed the level of permanent snow, the firn line and the tree line. At 2500 m at Sirunki in the New Guinea mountains the maximum late Pleistocene temperature depression, which occurred at 18 000 years BP, was about 10 °C (Fig. 5.1a), but at sea level it was only 2 or 3 °C. The various belts of montane rain forest were compressed as well as depressed; they came to occupy only three-quarters of their present area in three-fifths of their present altitudinal range. On present evidence the maximum depression of the tree line was 1500 m in New Guinea but only 350 m in Sumatra.

There have been continual fluctuations in the upper limit of forest and the tree line, shown for the last 30 000 years at Sirunki in Fig. 5.1b. The present day is atypical of the Quaternary as a whole, with sea level and the upper forest limit in the mountains exceptionally high, temperatures within 1 or 2 °C of their maximum Quaternary values and the climate probably at or near its wettest. Such conditions are believed to have existed for only a very small fraction of the whole Quaternary. Sea level may have risen at times to 3–7 m over present day levels but earlier reports that it has been much higher are now discounted (Haile 1971).

The conditions of the cooler, drier Glacial periods are believed to have affected the climate of Malesia in several ways (Verstappen 1975; Webster and Streten 1978). Firstly, the Intertropical Convergence Zone is likely to have been somewhat south of its present day roughly equatorial position. This is the zone where rising air of the

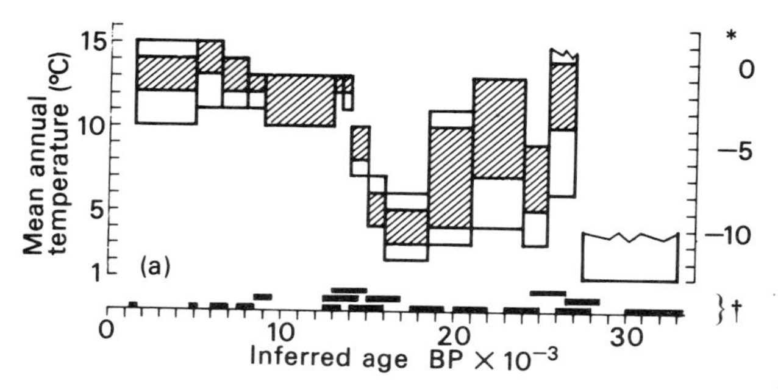

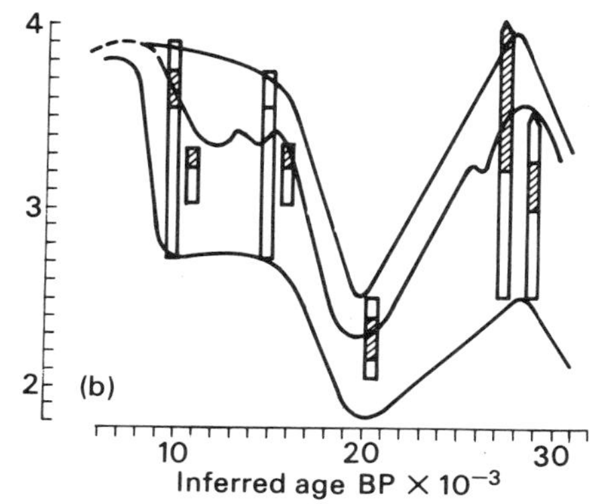

Fig. 5.1. Late Quaternary fluctuations in (a) temperature and (b) upper forest limit at Sirunki, 2500 m elevation in the New Guinea mountains.
* deviation from present-day mean annual temperature, °C.
† range of probable error in dating horizons.
The probable limits are shown hatched within the extremes and in (b) curves are drawn to join the different ages measured. (After Walker and Flenley 1979, Figs 14, 15.)

southern and northern hemispheres meets and its position in south east Asia is determined by the relative effects of the Asiatic and Australian anticyclones. It is a region of considerable showery rain and to its north is a high pressure belt of little rain which would have reached southwards into the archipelago when the Zone itself lay further south. The amount of southwards shift and consequential climatic change remains a matter of debate. But the effect would have been of a drier and more seasonal climate with lower rainfall and humidity and greater seasonal change in mean daily temperature.

Secondly, sea level 180 m below the present-day level would expose three times the present area of the Sunda continental shelf and twice the present area of the Sahul shelf (Fig. 2.1). Ocean currents, which today enter the archipelago mainly through the Torres Strait, the South China Sea, and south of Mindanao, would have been blocked, so their buffering effect on climate removed. The climate would have become more continental, with a greater diurnal temperature range and more valley winds which would have increased evapotranspiration. Rainfall and humidity would both have been lower than today.

Thirdly, the drop in sea temperature and lowland air temperature by a few degrees would also have reduced rainfall and humidity.

The sum effect would have been a climate with lower rainfall and humidity, of greater diurnal and seasonal fluctuations and with more marked rain shadows.

The present day seasonal climates, Fig. 5.2, would have been more extensive, and conversely the areas of perhumid climate would have been somewhat reduced.

Geomorphology

There are numerous and complex differences between perhumid and seasonally dry tropical climates in the form of weathering, erosion processes, deposited alluvium, and river channels. These are not yet fully explored for Malesia but many signs have been detected of the seasonally dry climates of the Pleistocene. For example, true laterite and kunkur nodules occur on the floor of the South China Sea, suggesting that, when this was exposed, it experienced a seasonal climate. In Malaya the Old Alluvium of Johore and Singapore and the alluvial deposits in the Kinta valley of Perak may be plains laid down when the climate was seasonal. Furthermore, the steepland boundary in Malaya, where the gently graded plains abruptly meet the sharp relief of the mountains, may also result from erosion and deposition in a seasonal climate.

Future geomorphological research might help elucidate more precisely the past distribution of seasonal climates, in the way it has in the Amazon basin (Haffer, in preparation).

QUATERNARY VEGETATION HISTORY

The two rain forest blocks

The tropical rain forest of Malesia today occurs in two great blocks, Sundaland and Papuasia, centred respectively on areas of perhumid, aseasonal climate in Sumatra, Malaya and Borneo, and New Guinea, and separated by a north to south corridor of seasonally dry climates running down the western side of the Philippines, through Celebes and the Moluccas to the Lesser Sunda Islands and Banda Arcs, and then extending west to Java, where it narrows as a wedge along the north coast, and

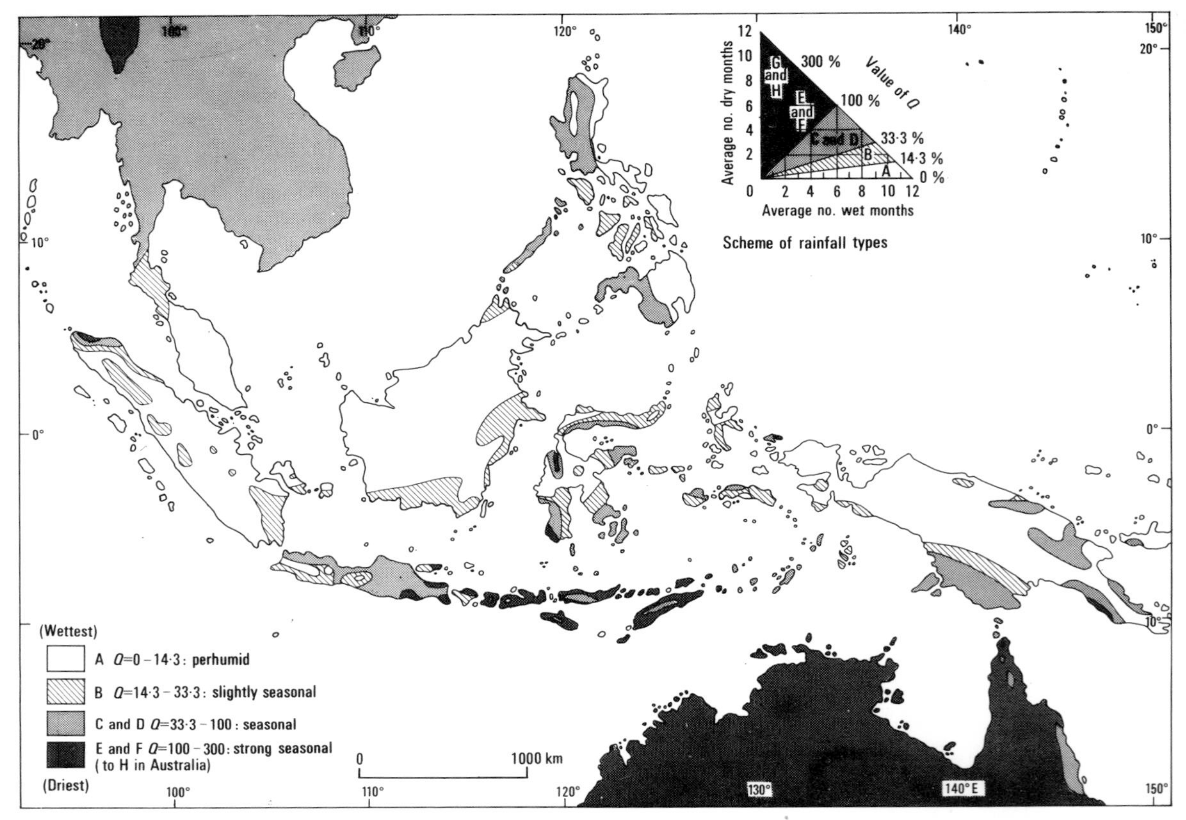

Fig. 5.2. Rainfall types in Malesia based on dry/wet period ratios. This climatic index, Q, developed by Schmidt and Ferguson (1951) is the only one which closely reflects present-day distributions of vegetation types. Based on Whitmore (1975, Fig. 3.1, corrected), where further details are given. Q is defined as (dry months/wet months) × 100 where wet months are those with over 100 mm precipitation and dry those with less than 60 mm.

east to encompass the southernmost part of New Guinea (Fig. 5.2; and Whitmore 1975). In the seasonal areas there occurs a complex mosaic of rain and monsoon forest and the original pattern has been confused by the hand of man. In areas of seasonal climate rain forest persists in wetter sites, especially as gallery forest along rivers, and also on the better soils.

In the drier, more seasonal periods of the Pleistocene the areas of wet, perhumid climate would have contracted as the area of seasonal climate extended. The exact amount of change remains imponderable. Populations of rain forest species would have had their ranges constricted, but contact could have been maintained to various indeterminate degrees via the pockets and strips of rain forest in the seasonal areas, as occurs today for example in South America between the Atlantic coast and Amazonian rain forests (Prance, in preparation). The patches of rain forest in east Java today are perhaps representative.

It is likely that monsoon forest, and perhaps savannah forest too in the driest places, would have occupied more of the Philippines, Celebes, and eastern and southern Borneo and a greater area of Sumatra, Java, and New Guinea (Fig. 5.2).

There is, however, no evidence, from the few lowland pollen profiles which have been constructed within Malesia, of any replacement of rain forest by monsoon forest during or since the Tertiary (Morley 1977; J. Muller, personal communication) but this may be supposed simply to reflect the inadequate sample. In one area of north Queensland, which lies on the present day boundary of rain and monsoon forest, the pollen record does show strong climatic fluctuation during the Quaternary.

Monsoon climate plant species

Plant species of monsoon climate which occur within Malesia are of mainly Asian origin, a few are Australian and a very small number are endemic, several having a very wide range from India, or even Africa, to Australia (van Steenis 1957, 1979; and briefly in Whitmore 1975). Within the Malay archipelago their distribution is disjunct, as isolated populations between which dispersal is improbable. The fact that there are very few endemics and very few differences between the isolated populations is evidence that monsoon plants have had more nearly continuous ranges in the past, i.e. that there have been more extensive areas of seasonal climate (van Meeuwen, Nooteboom and van Steenis 1961). Figure 5.3 shows two

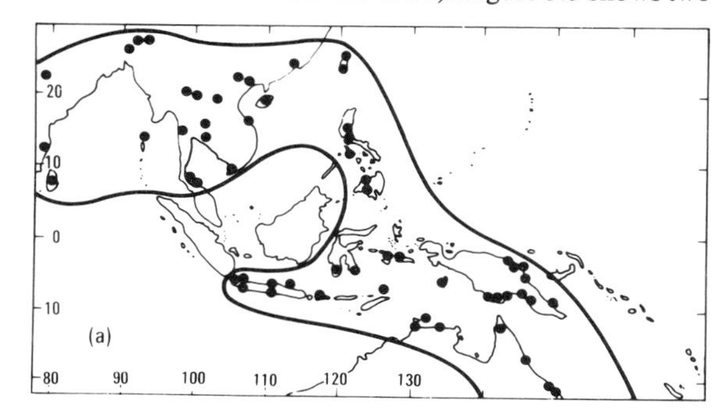

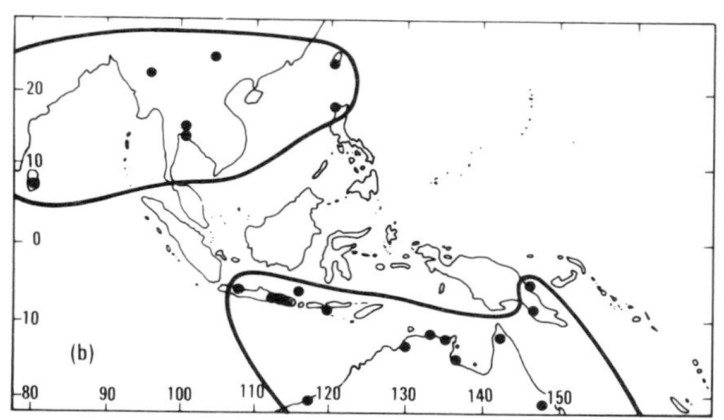

Fig. 5.3. Geographical ranges of two monsoon climate trees of the family Papilionaceae, (a) *Pycnospora lutescens* requiring some degree of dry season and (b) *Rhynchosia minima*, requiring a much stronger dry season: the disjunction is much more marked (van Meuwen, Nooteboom, and van Steenis 1961).

examples of trees of the Papilionaceae whose degree of present day disjunction is a measure of how dry a climate they require.

Mountain plants

Mountains occur as 'islands' surrounded by a 'sea' of lowland. The mountain floras of Malesia have been fully and critically examined by van Steenis (1934, 1936, 1962, 1965, 1972*a*; very briefly summarized by Whitmore 1975 and Flenley 1979). Three main paths of entry and ranges within the region can be recognized as the Sumatran, Taiwan–Luzon, and New Guinea tracks (Fig. 5.4). The high mountain plants are of either north or south temperate affinity, for example *Primula prolifera*, which reaches east Java along the Sumatran track, and the orchid *Caladenia carnea* along the New Guinea track. They could not have reached their present extension under present day climatic conditions. But during the Quaternary, when at the coolest periods treeline was depressed by up to 1500 m and montane forest belts were also depressed, there would have been greatly extended areas of suitably cool habitat. The stepping stones required for dispersal would have been very much closer.

Pleistocene rain forest refugia

Within the tropical rain forests of both West Africa (White 1978) and tropical America (Prance 1981; Whitmore, in preparation) there have now been detected marked centres of species richness and endemism for plants and animals. These are believed to represent refugia to which the rain forest retreated at the cool dry periods of the Glacial maxima. The refugia were 'islands' of rain forests, some of them centred on mountains, and in a 'sea' of monsoon and savannah forest.

In the Malay archipelago the Sundaland and Papuasian rain forest blocks are comparable to these refugia in terms of species richness and high degree of endemism. Some of the proposed refugia in the Amazon basin are very large (Fig. 5.5). There is no evidence that the Malesian rain forest was ever reduced to isolated 'islands' in a large savannah 'sea' although the evidence from monsoon climate plants, landforms and mountain plants is that lowland rain forest in the Malay archipelago was, as elsewhere, of reduced extent. Here, as elsewhere, Pleistocene desiccation did cause its contraction.

The lower sea level exposed large areas of the continental shelves of low relief. Parts would probably have had saline soils and borne extensive mangrove and brackish water forests but away from saline influence we do not know how much bore inland rain forests and how much bore seasonal forests. We can speculate that here, on land currently below as much as 180 m of sea, there developed in the Far East big expanses of monsoon and savannah forests comparable to those now known to have occurred in Africa and America. The Malesian part of the tropical rain forest belt differs from the African and American in being much more mountainous. Both the equatorial part of the Andes and the mountains of Cameroun are close to important rain forest refugia. As the climate became wetter rain forest spread out from

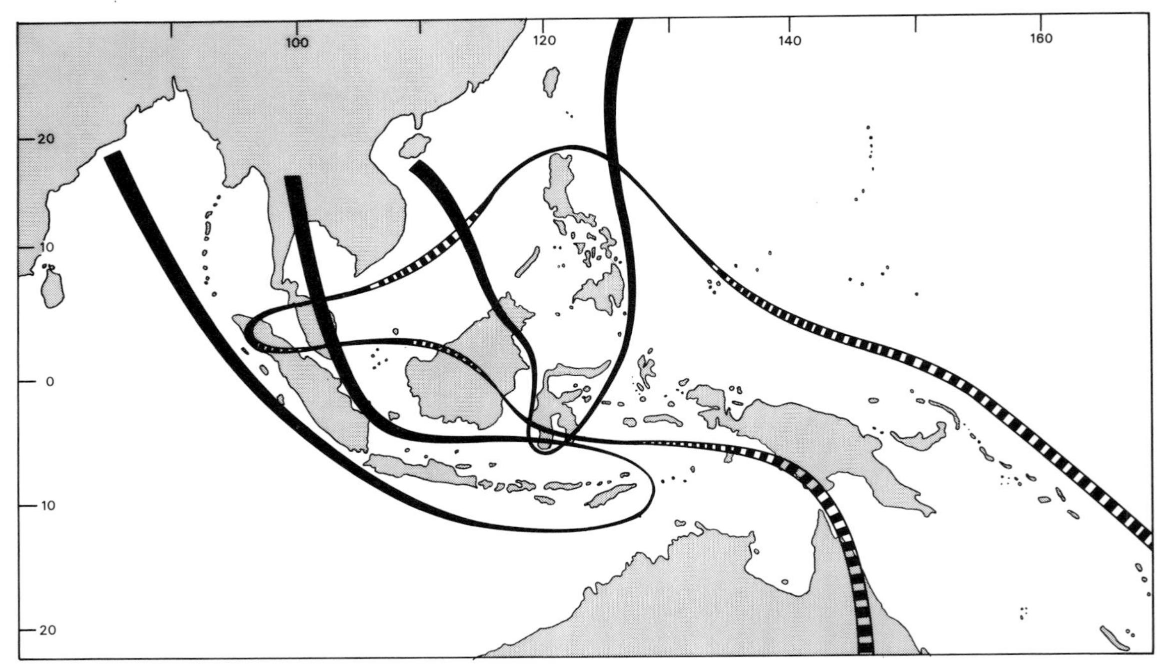

Fig. 5.4. The Sumatran, Taiwan–Luzon, and New Guinea tracks of entry of mountain plants to Malesia entering from north-west, north, and south-east respectively (van Steenis 1965, Fig. 11).

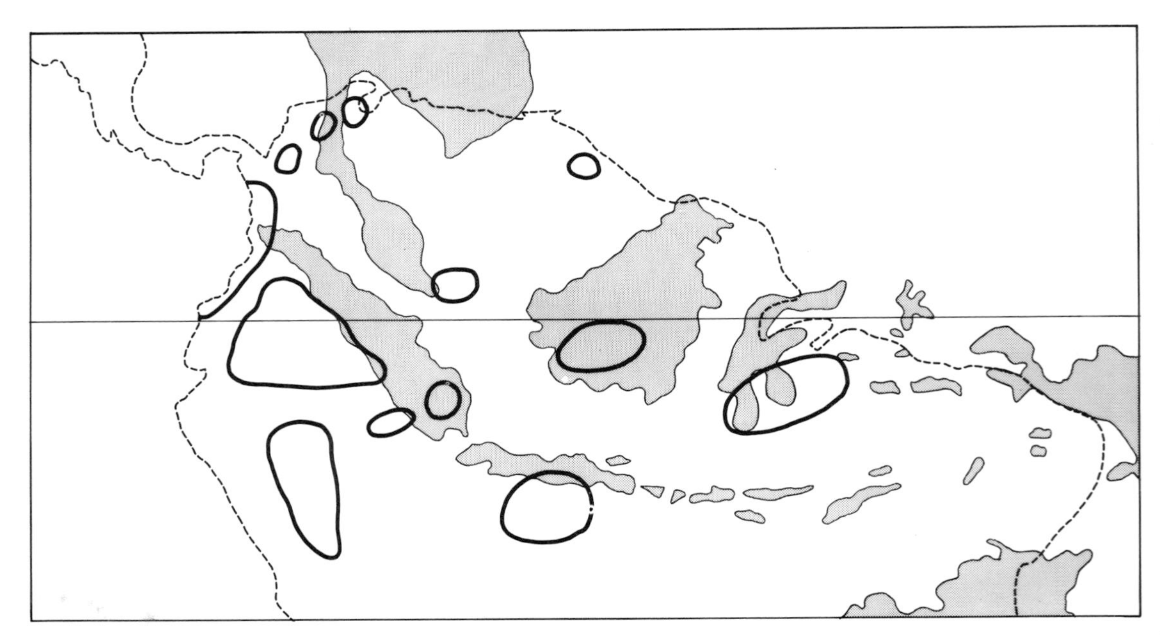

Fig. 5.5. Northern South America with part of Malesia superimposed. Neotropical Pleistocene rain forest 'refugia', postulated from ranges of members of four tree families (Prance 1973), are shown.

these, but in Malesia it is suggested that the concomitant rise in sea level drowned the seasonal forests which surrounded the mountainous rain forest refugia thus giving rise to the differences we see today.

Lowland rain forest plants and animals

At the present day most of Malesia is clothed by rain forest (which man is fast decimating) and it is in the ranges of rain forest plants and animals that we can expect to see the main signs of distribution patterns which result from the collision of Laurasia and Gondwanaland 15 Ma ago, once the complications caused by Pleistocene climatic change have been interpreted. To this investigation we shall turn our attention, in Chapters 6, 7, and 8. But before doing so we end this chapter with a speculative look at Tertiary vegetation history.

TERTIARY VEGETATION HISTORY

Besides the changes in vegetation considered so far in this chapter and which resulted largely from fluctuating Pleistocene climate there must have been others which followed the late Tertiary creation of Malesia. These we can only perceive in broad terms and with a degree of speculation. This longer perspective, even if hazy, does however, help to focus on matters which now require closer study.

The mid-Miocene collision which created the Malay archipelago and the subsequent emergence of a close chain of islands during the late Miocene to late Pliocene (Chapters 3 and 4, Table 4.1) enabled two floras to intermingle. The more easily dispersed plants would have been able to cross the diminishing sea gap before actual collision. The migration of individual genera through the archipelago is traced for palms and some other plants in Chapters 6 and 8 and evidence of evolutionary relationships and worldwide ranges is brought to bear. Sound taxonomy, based on numerous collections from the whole geographical range of a group, is an essential prerequisite for this kind of analysis.

The genera chosen for discussion are believed to be sufficiently well known for us to conjecture their past history from the pattern perceived today. In fact, considering the whole Malesian flora, present day ranges suggest there are very many plants with Laurasian origin and rather fewer with Gondwanic origin. This may be deceptive as little is yet known about rates of speciation and extinction in the 10 Ma or so which have elapsed since massive migration became possible.

There have been considerable fluctuations in climate and vegetation and in geology (for example Mount Kinabalu, the highest peak between the Himalayas and New Guinea, is only 1·5 Ma old) whose effect on plant speciation is unknown.

But if present plant ranges are indeed indicative of their past history it is interesting to consider the floristic composition of different vegetation types. Does this too mirror palaeogeography?

We observe that in the rain forests of New Guinea Laurasian floristic elements are strongly predominant, reflecting a massive invasion of this northern fringe of Gondwanaland from the west. By contrast to the greater part of the New Guinea rain forests those lower montane rain forests in which *Nothofagus* is common and *Araucaria* locally dominant may perhaps be essentially Gondwanic, their flora evolved in a moist temperate climate, rafted northwards and persisting in the wet tropical montane environment. Floristically similar vegetation now occurs disjunctly from New South Wales southwards, but has disappeared from Queensland following Australia's desiccation (Chapter 8; Walker 1972).

West of Wallace's line there is no comparable massive influx into lowland mesic rain forest from the east. The Gondwanic (Australian) floristic element has invaded Laurasian Malesia but is found as a major component only of the heath forest flora. This floristic interdigitation was identified by Richards (1943) and analysed in some detail by van Steenis (1979), who argued that the Australian flora became specialized to oligotrophic conditions and only succeeds in Malesia where these occur.

Seasonally dry Malesia, clothed with monsoon and savanna forests, has a greater Gondwanic floristic component than the rain forests. The Lesser Sunda Islands, which are all seasonally dry, contain a substantial Gondwanic flora which diminishes westwards (Kalkman 1955). The great southern bulge of New Guinea itself also bears monsoon and savannah forests with a markedly Gondwanic flora, including several *Eucalyptus* species shared with northernmost Australia.

We have as yet little detailed, critical knowledge of the floristic composition of different sorts of forest for New Guinea and other parts of Gondwanic Malesia. It is yet to be discovered whether there are enclaves of Gondwanic and Laurasian flora as clear-cut as the heath forest/mesic rain forest dichotomy of Laurasian Malesia west of Wallace's line. New Guinea is mountainous and climatic changes during the last 2 Ma, the Pleistocene, have caused repeated fluctuations in altitude of forest types and tree species. It has even been suggested that on a mountain different species migrate at different rates in response to the same change in temperature so that, at the recently achieved present-day climatic optimum, we see montane forests still recruiting species from lower elevation (Walker and Flenley 1979). That is to say that, although the forest types, defined on structure and physiognomy, which are a response to environment, form a regular repeating pattern, the species vary from place to place, dependent on the vagaries of migration and chance local extinction.

The great fluctuations in climate and changes in geology and the enormous length of time relative to the life-span of even the longest lived tropical plant since Malesia was created, inevitably confuse detection of simple patterns. Plate tectonic history is but one factor amongst many others. Detailed knowledge, which will only come by painstaking further research in these forests, is needed to fill out and validate the broad, speculative picture sketched here. We can see, *inter alia*, the substantial intellectual loss if the primeval forest is eliminated in a late twentieth century rush to 'mine' it for timber.

6 PALMS AND WALLACE'S LINE

J. Dransfield

The palms are a large, complex, and mainly tropical family, especially richly developed in the Malay Archipelago. Dispersal ability is generally low. There are many local endemics. The modern view is that there are several different natural alliances within the family and there is evidence for their origin in West Gondwanaland, (present day South America). Within Malesia a majority of genera is centred on Sundaland, and a few on Papuasia. From both centres there are outliers in the other whose origin could be due to migration since the Miocene collision of Laurasia with the Australia/New Guinea shard of Gondwanaland. Within Celebes Gronophyllum, *mainly Papuasian, is known only from the eastern side. Several genera and alliances have a concentration in both Sundaland and Papuasia. When their relatives elsewhere are also considered this pattern is consonant with origin from ancestors in unfragmented Gondwanaland which moved northwards on both the India and Australia/New Guinea fragments to reach Malesia twice, from both the north-west and south-east. Finally, a few genera have ranges which defy interpretation on present geological knowledge.*

INTRODUCTION

The palm family (Palmae or Arecaceae) is peculiarly well suited to distributional analysis in the Malesian region. It is a large, apparently ancient, and very diverse family of flowering plants mostly confined to and epitomizing the tropics and subtropics, and is particularly well developed in Malesia. Furthermore, and perhaps most importantly, the family has recently been the subject of a detailed reassessment at the generic and subfamily level (Moore 1973*b*). The results of the reassessment are embodied in a totally novel rearrangement of palm genera into 'Major Groups', in many instances quite different from previous classifications; Moore's analysis is particularly convincing as it incorporates information from most recent palm research. The palms are also probably much better indicators of past geography than are some families, because of the often large size of their fruits, which suggests long-distance dispersal might be limited. Rhododendrons and orchids might, on the other hand, be expected to be much more widespread at the generic and sectional level because of the very small seed, usually dispersed by wind, allowing for perhaps much quicker colonization of newly juxtaposed land masses. Although the generic taxonomy of the palms is now relatively well understood, there are still major lacunae in our knowledge of specific distribution, especially in such large genera as *Calamus*, *Daemonorops*, and *Licuala*; some areas of Malesia, such as the Moluccas, are still poorly collected, and in particular the rattan flora of Celebes is still only partially represented in herbaria. (Rattans are the climbing palms used for cane furniture.) Despite this, Moore's analysis provides an excellent basis for discussion of the distribution of the family across Wallace's line and through Malesia; this account could not have been written without the basis of Moore's paper. Fossil palm material is, as might be expected, sparse and has not been recorded for the really crucial area, Celebes; yet what fossil material has been recorded, is of very great interest. In particular the palynological sequences from Borneo recorded by Muller (1968) and Anderson and Muller (1975), lend evidence which certainly does not contradict the general theses from present day palm distributions and historical geology.

The palm family comprises about 2700 species arranged in 212 genera which can be arranged in fifteen Major Groups, representing perhaps five distinct lines of evolution (Moore 1973*b*). The differences between these major lines of evolution are great, often much greater than those frequently used to delimit families of other angiosperms (e.g. the families within the order Scitaminae which include bananas, gingers, and cannas); despite this, the order Principes with its one family, the palms, represents a very natural and ancient order. Within the major groups Moore recognizes natural alliances of genera. Many of the major groups and alliances have wide geographical ranges. But in general the family displays a rather high degree of endemicity at both the generic and specific level, although there are also a few genera of very wide distribution. For example, *Calamus* is distributed from West Africa to Fiji and Australia and *Phoenix* (the date palm and allies) is known from the west Atlantic islands through Africa to China and south-east Asia. Another example is *Rhopaloblaste*, discussed below. Such widespread genera are, however, exceptional; most are confined to relatively small areas, and some floras consist of numerous monotypic or small endemic genera for example the palm flora of New Caledonia. The relationships of such highly endemic floras would indeed be difficult to analyse were it not for Moore's arrangement into Major Groups, alliances, and units.

Within the botanical province of Malesia, palms form an important component in most forest types, occurring from sea-level up to altitudes of 2800 m or more in the mountains. One growth form, the climbing palms including the rattans, is peculiarly well-developed throughout the region, though rattans are also present in Africa and some unrelated climbing palms are sparsely represented in South America. The rattans frequently comprise half or more of the palm flora of a given locality in Malesia.

Despite the traditional treatment of the Malesian region as a botanical province, and the superficially similar role that palms play in the forests of the whole region, the palm flora of the Sunda Shelf lands is strikingly different from that of the Gondwanic islands of Papuasia; though there is a transitional zone between the two areas, the general impression is one of disjunctions.

If the two floras are so distinctive, where and how do they impinge on each other, and where are their origins?

The origin of the palms as a whole has been the subject of much speculation; until recently, an origin in the perhumid tropics of what is now south-east Asia has been rather uncritically proposed – the so-called 'cradle of the Angiosperms' (Takhtajhan 1969). Moore (1973*a*), using his analysis of the Major Groups, has proposed a quite different origin. Data from present day distribution patterns, morphological analyses, an interpretation of what is probably primitive and what advanced, and the few palm fossils available, suggest an origin of the palms in West Gondwanaland, i.e. what is now South America, followed by dispersal in several directions. In his paper Moore even hinted that the palm flora of New Guinea might have been partially recruited via an austral route, the presence of fossil palm pollen in Antarctica being evidence for this suggestion. The new geological evidence makes this theory highly plausible; in fact, the putative West Gondwanaland origin for the family will be accepted as a basis for discussion in the present chapter.

FOSSIL PALMS

Palm fossils are generally scarce, as might be expected. The few relevant fossils of present day Malesian genera suggest a more widespread distribution for some genera than at present. Moore (1973*b*) and Muller (1968; 1972) give valuable summaries of the known fossils. The most remarkable is the genus *Nypa*, which is known as fossil pollen from the Upper Cretaceous of Borneo till the present day, with macrofossils in the Eocene and Miocene of Europe, Africa, and India and the Palaeocene of Brazil (Tralau 1964). *Caryota* seeds have been recorded from the London Clay (Eocene) and pollen from the Isle of Wight (Oligocene) (Reid and Chandler 1933; Machin 1971) and from the lower Eocene of India (Lakhanpal 1970). *Nenga* pollen is known from the mid-Miocene of Borneo (Muller 1972). In New Zealand, pollen of *Rhopalostylis*, the only known native palm, is first found in the Miocene, and fruits and pollen of Cocosoid palms are also recorded (see Moore 1973*b*).

Scanty though this fossil record is it demonstrates that some palm genera appear to have a long geological history. Similar discoveries have been made for many other higher plants. The implication is that evolution at the generic level is slow. This becomes an important basis for the discussions in this book about past distributions of plants (this chapter and Chapter 8).

THE PALM FLORAS OF WEST AND EAST MALESIA COMPARED

The differences between the palm floras of west and east Malesia are shown in Table 5.1 illustrating the distribution of genera and approximate number of species in different parts of the region. The Philippines are not included in the table; two genera, *Heterospathe* and *Veitchia* are present in the Philippines and New Guinea and the Philppines and the west Pacific respectively and will be mentioned again later. Nineteen genera of Sundaland (Table 6.1 a–c) are not found in New Guinea. Conversely, at least twelve genera found in New Guinea are not found west of the Moluccas (Table 6.1 d) and a further genus (*Pigafetta*) is not found west of Celebes (Table 6.1 e). There is thus a big difference between the two ends of Malesia. About eighteen genera are found in both areas (Table 6.1 f) but most of these are unevenly distributed with the greater representation being in either the west or the east. Most of the shared genera have a greater representation and diversity in the west and decrease in representation eastwards until there may only be one or two species in Papuasia. *Arenga*, *Daemonorops*, *Korthalsia*, and *Pinanga* are good examples of such genera. We may regard them as being genera originating in Sundaland or further west and only recently having reached Papuasia, where little if any speciation has taken place. There is further evidence in the instance of *Arenga* where Muller (1972) records fossil pollen in the mid-Miocene of Borneo, that is, before the collision of Laurasia with Gondwanaland.

The major disjunction in the palm flora can be seen to be east of Borneo and west of New Guinea, i.e. in the area of the Moluccas and Celebes. We still know relatively

Table 6.1

Geographical ranges of genera of palms within Malesia and the numbers of species

(a) **Eastern limit in Malay Peninsula**

	Species in Sundaland
Maxburretia	2
Phoenix	1
Myrialepis	1
Calospatha	1

(b) **Eastern limit in Borneo**

	Species in Sundaland
Iguanura	21
Salacca	17
Plectocomia	7
Ceratolobus	6
Eugeissona	5
Johannesteijsmannia	4
Plectocomiopsis	4
Nenga	4
Borassodendron	2
Pichisermollia	2
Pogonotium	2
Eleiodoxa	1
Retispatha	1

(c) **Eastern limit in Celebes**

	Species in	
	Sundaland	Celebes
Pholidocarpus	4	1
Oncosperma (possibly also in Moluccas)	2	1

(d) **Western limit in Moluccas**

	Species in Papuasia
Linospadix	26
Ptychosperma	23
Nengella	19
Calyptrocalyx	17
***Heterospathe*	16
Hydriastele	7
Gulubia	6
Ptychococcus	6
Drymophloeus	4
Sommieria	3
Brassiophoenix	2
Siphokentia	1

(e) **Western limit Celebes**

	Species in	
	Celebes	Papuasia
Pigafetta	1	1

(f) **Throughout Malesia**

	Species in		
	Sundaland	Celebes	New Guinea
Calamus	120	21	50
Daemonorops	70	6	?1
Pinanga	70	5	1
Licuala	*c.* 50	2	36
Korthalsia	20	1	1
Areca	15	5	11
Arenga	12	2	1
Livistona	6	1	6
Caryota	3	2	1
Orania	1 (?2)	?1	8
† *Cyrtostachys*	1	–	8
† *Rhopaloblaste*	1	–	4
Corypha	1	1	1
Nypa	1	1	1
Gronophyllum	?1	4	7
Actinorhytis	1*	1*	1
Borassus	1*	1?*	1
Metroxylon	1*	1*	1

† Not in Celebes
* Introduced
** Also in Philippines

little of the palm flora of the Moluccas, but there are several Papuasian genera such as *Calyptrocalyx*, *Drymophloeus*, *Gulubia*, *Ptychosperma*, and *Rhopaloblaste* which are found there, suggesting a much more diverse flora at the generic level than that of Celebes.

The striking difference between the Papuasian and Sundaland palm floras fits in well with the geological evidence of only relatively recent (i.e. Miocene) juxtaposition. There are, however, as mentioned above, many genera with striking diversity on both the Laurasian and Australia/New Guinea plates, and there are some genera present on both plates but absent from Celebes. An examination of the palm flora of Celebes is thus necessary.

THE PALM FLORA OF CELEBES

The palm flora of Celebes itself, though still not particularly well-known, is generally a poor one. So far, only fifty-two obviously indigenous species in fifteen genera have been recorded, Table 6.2. None of the genera is endemic. At the specific level there is also a relatively low degree of endemism except in the rattans *Calamus* and *Daemonorops*.

It is known that the climate of Malesia was drier during the Pleistocene and Celebes appears to be one of the areas to have been most affected (see Chapter 5). Thus a drier Pleistocene climate may be one of the reasons for the paucity of palms in Celebes.

Of the fifteen genera the strongest affinity is with Sundaland. Only two genera are not found west of the Makassar Strait: *Pigafetta* (Fig. 6.1), monotypic, found from New Guinea, Moluccas, and Celebes and *Gronophyllum* (Fig. 6.2), a genus present in north Australia, New Guinea, and Celebes. *Pigafetta filaris*, a huge tree palm reaching great heights, is a pioneer species,

Table 6.2

The palm flora of Celebes

	Endemic species	*Non-endemic species*
Calamus	*c.* 15	6
Daemonorops	*c.* 5	1
Gronophyllum	4	–
Pinanga	4	1
Areca	*c.* 3	2
Licuala	1 (?2)	–
Korthalsia	1	–
Arenga	–	2
Caryota	–	2
Corypha	–	1
Pigafetta	–	1
Nypa	–	1
Livistona	–	1
Pholidocarpus	–	1
Oncosperma	–	1
Actinorhytis	–	1*
Borassus	–	1*
Metroxylon	–	1*
Salacca	–	1*
Orania†	?	

* Introduced
† Based on photograph only (Fairchild 1943)

found on landslips and in seral forest in the mountains (Dransfield 1976, 1978); it has relatively small seeds, produced in great abundance and it is thus all the more remarkable that it has not crossed the Makassar Strait to Borneo. *Gronophyllum* has fifteen recorded species, four of which are described as Celebesian endemics; however, the genus is greatly in need of revision and it is possible that all four Celebes taxa are conspecific. Within Celebes, *Gronophyllum* is as yet only known from the south-east arm and from the east-central area by Malili and the Poso Lakes, and this confinement to the Gondwanic fragment may be significant. There are two anomalies in *Gronophyllum* which require investigation. There are two trees of a *Gronophyllum* in cultivation in Semongok Arboretum near Kuching, Sarawak, said to have originated from seed collected in a peat swamp at Marudi in the 4th Division, Sarawak; there are no wild source herbarium specimens, and the species (as yet unknown) has been seen nowhere else in Borneo and there could well have been a mixing of labels in the nursery at the seedling stage, between a Bornean palm and *Gronophyllum* from, perhaps, New Guinea. The Sarawak plant collector Jugah ak Kudi is, however, insistent that the *Gronophyllum* originated in Sarawak. Similarly there is one tree of an unknown species of *Gronophyllum* in cultivation in the private garden of Professor J. Pancho in Los Banos, Luzon, said to have come from seed collected in Mindanao; if mixing of labels has not occurred, the genus could thus also be present in the Philippines.

DISJUNCTIONS AND BICENTRIC DISTRIBUTION PATTERNS

More difficult to account for are genera found on either side of Wallace's line but absent from Celebes itself, or those genera such as *Livistona* and *Licuala* (Table 6.1 f)

Fig. 6.1. *Pigafetta filaris*, a monotypic Lepidocaryoid palm of New Guinea, Moluccas, and Celebes.

which are relatively poorly represented in Celebes but have large and diverse representation on either side. With the collision of the Laurasian and Australia/New Guinean plates during the Miocene, the presence of one or two species of a largely Asiatic genus such as *Korthalsia* in New Guinea can be conveniently explained as due to Miocene or post Miocene migration, but much

Fig. 6.2. *Gronophyllum selebicum*, near Malili on the Gondwanic part of Celebes.

more complex explanations are required for the genera like *Licuala*, and *Livistona*.

Cyrtostachys

Cyrtostachys, a very natural though isolated genus (Moore 1973*b*), consists of seven species in New Guinea and the Solomon Islands and one species in Sumatra, the Malay Peninsula (including South Thailand) and Borneo, Fig. 6.3. The Sundaland species, *C. renda* (the Sealing Wax Palm, usually known as *C. lakka*) is a plant of peatswamp forest and is mainly coastal. This habitat is normally one with low palm endemism—the peatswamp palm flora is relatively similar in species throughout Sundaland. The diversity of the genus east of Wallace's line is suggestive of its spread westwards to Sundaland rather than eastwards. *Cyrtostachys* is one of the few genera with a fossil record in Malesia and the first fossil record in Sundaland is in Borneo in the Upper Miocene (Muller 1972). If the genus originated in Australia/New Guinea as suggested here, then it appears to have reached Sundaland soon after the collision of the plates. It is indeed remarkable that there are no vestiges on its putative migration route through the Moluccas and Celebes, though this might be due to extinction, perhaps due to the absence at the present day of peatswamp forests there, or to the presence of drier conditions during the Pleistocene.

Rhopaloblaste and the Clinostigma alliance

Rhopaloblaste, another well circumscribed genus, has a distribution disjunct across Wallace's line but much less easy to explain than that of *Cyrtostachys* (Fig. 6.4). There are seven species (Moore 1970). *R. augusta* is confined to the Nicobar Islands and *R. singaporensis* to the Malay Peninsula (including Singapore). Then there is a gap until the Moluccas with *R. ceramica*, which is possibly also in New Guinea (Essig 1977) where *R. brassii*, *R. dyscrita*, and *R. ledermanniana* also occur. Finally, *R. elegans* occurs in the Solomon Islands. There is no fossil record. Thus, the genus is bicentric within Malesia. Its distribution might perhaps result from an origin on either the Australia/New Guinean or Laurasian plate, followed by migration after the collision. But this supposes total

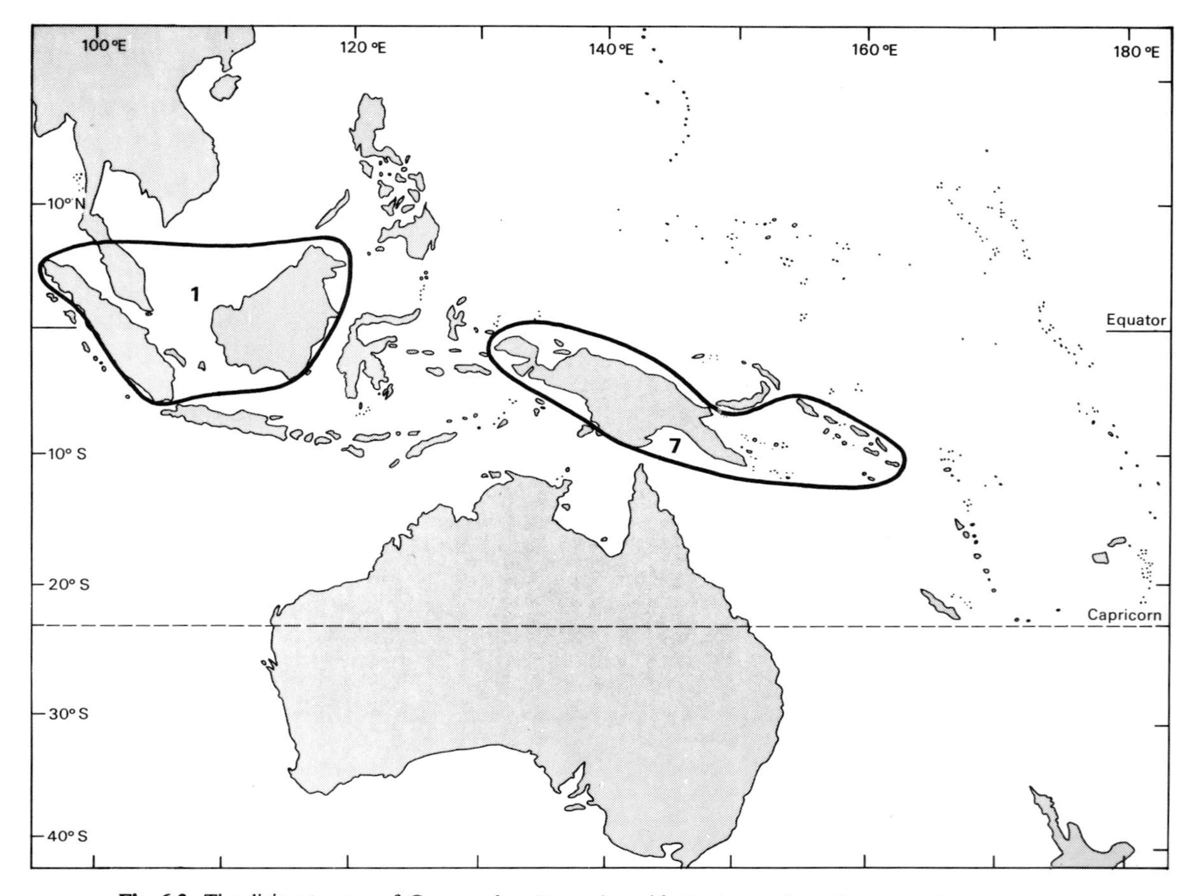

Fig. 6.3. The disjunct range of *Cyrtostachys*, Papuasian with Sundaic outliers. Numbers of species shown.

extinction in the whole region between Malaya and the Moluccas, to account for which drier conditions during the Pleistocene could be invoked. There is another possibility (already hinted at by Moore 1973*a*) which is that *Rhopaloblaste* reached present day Malesia from both the west and the east. *Rhopaloblaste* is a member of the *Clinostigma* alliance of genera (Fig. 6.5). Most members are found in Papuasia, Fiji, Samoa, and New Caledonia, all of which are parts of the Australia/New Guinea fragment of Gondwanaland. But well to the west of Wallace's line there is a second, though lesser, centre of diversity represented by *Bentinckia*, a genus of two species, one of southern India (*B. condapanna*), one of the Nicobars (*B. nicobarica*) and *Dictyosperma*, monotypic (H. E. Moore, personal communication) endemic to the Mascarene Islands in the southern Indian Ocean. The possibility exists that the *Clinostigma* alliance existed on old Gondwanaland, and members were carried northwards both on the Indian and Australia/New Guinea plates when Gondwanaland broke up. *Rhopaloblaste* itself could have been on both fragments and have entered present day Malesia and the Nicobars from the north via India as well as from the south.

Iguanura, eighteen species, was considered to belong to this generic alliance by Moore (1973*b*), though this has been disputed by Kiew (1976). It is a highly characteristic undergrowth palm genus of the rain forest of Sumatra, the Malay Peninsula (including south Thailand) and Borneo, with remarkable diversity in Borneo. Traditionally it is believed similar to *Sommieria*, a New Guinea endemic genus with three species (Beccari 1877). The marked geographical disjunction between *Sommieria* and *Iguanura* could suggest that similarities in morphology are due to parallelism; but *Iguanura* and *Sommieria*, like *Rhopaloblaste* may have a common

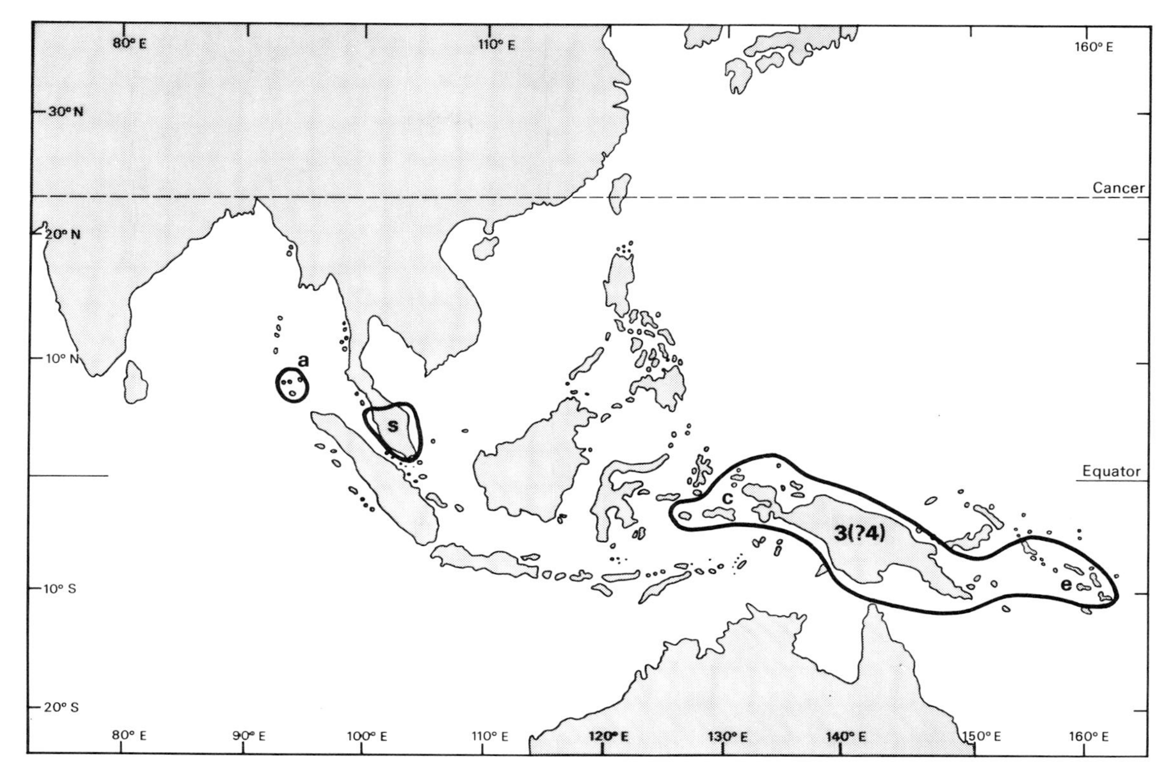

Fig. 6.4. The disjunct range of *Rhopaloblaste*. **a**: *R. augusta* (Nicobars); **c**: *R. ceramica*; **e**: *R. elegans*; **s**: *R. singaporensis*.

Gondwanic ancestor. Relationships between the genera are much in need of careful reassessment.

Other Clinostigmatoid genera are to be found on the Western Pacific islands (Fig. 6.5). *Heterospathe* is found in New Guinea, the Solomon Islands, Moluccas, Philippines (where there is a minor radiation) and Guam at the southern end of the Marianas archipelago. *Satakentia* with one species only (*S. liukiuensis*) is known only in the Ryukyu islands (Moore 1969*a*). *Clinostigma* itself, with species in New Ireland in the Bismarck archipelago, the Solomon Islands, New Hebrides, Fiji, and Samoa has also species in the Caroline and Bonin islands (Moore 1969*b*). These distributions suggest dispersal outwards from the Australia/New Guinea plate by the migration routes into the Pacific described in Chapter 4.

The Lepidocaryoid palms

The Lepidocaryoid Major Group reaches its greatest diversity in Malesia, though it is represented in New and Old Worlds, and from Africa through to Fiji, Fig. 6.6. Moore (1973*a*) regarded the African representatives *Ancistrophyllum*, *Eremospatha*, *Oncocalamus*, and *Raphia* as being some of the simplest and possibly most primitive in the Major Group. In the Americas *Raphia* is also present (one species) and two extraordinary fan-leaved genera, *Lepidocaryum* and *Mauritia*, (all other Lepidocaryoids are pinnate-leaved) complete the representation. *Calamus* stretches from Africa to Fiji, but all the other genera are south-east Asian or Malesian (Fig. 6.7). Van der Hammen (1957) recorded the presence of *Mauritia* pollen in the Palaeocene of South America, arguing a long history of the genus there, and rattan pollen occurs in the Palaeocene of Borneo (where it is very abundant) (Muller 1968) and the Eocene of India (Mathur 1963) and Europe (Gruas-Cavagnetto 1976). There is no doubt, then, that the rattans have undergone their major development west of Wallace's line and, as indicated above, all genera are very poorly represented east of the line except *Calamus*. Could *Calamus*, too, have entered Malesia from the south-east as well as from the north-west, assuming an African origin of the Major Group? In some ways this seems unlikely. *Calamus* is the largest rattan genus and displays a great radiation of species in each island, even in Celebes, suggesting,

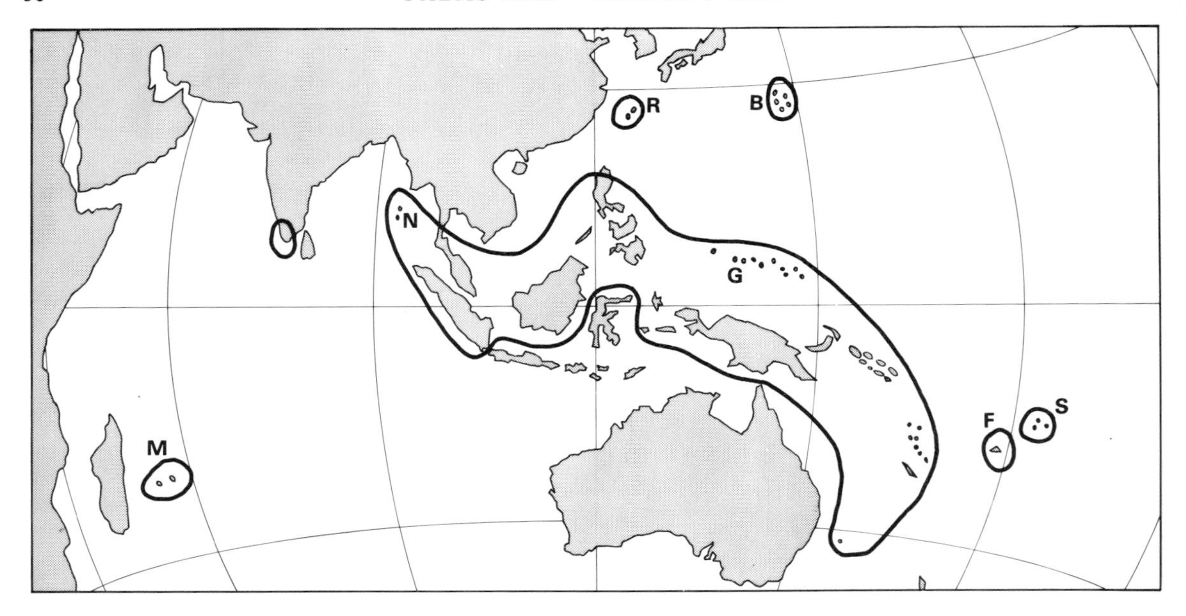

Fig. 6.5. The range of the *Clinostigma* alliance of genera. **B**: Bonin islands; **F**: Fiji; **G**: Guam; **M**: Mascarenes; **N**: Nicobars; **R**: Ryukyu islands; **S**: Samoa. Lambert equal area projection.

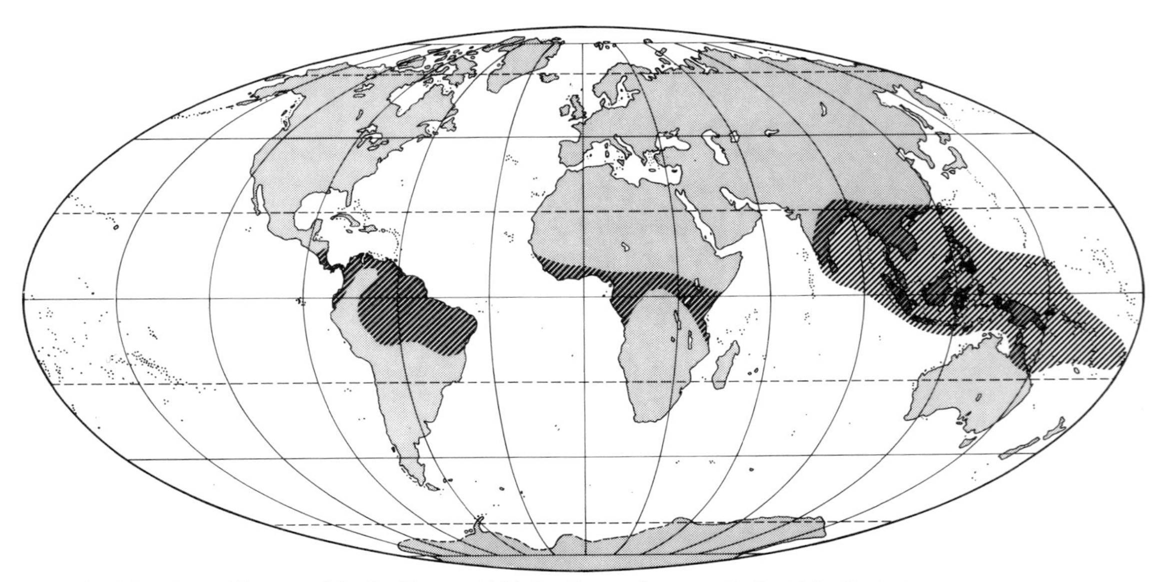

Fig. 6.6. The world range of the Lepidocaryoid Major Group of genera. Mollweide's elliptical equal area projection.

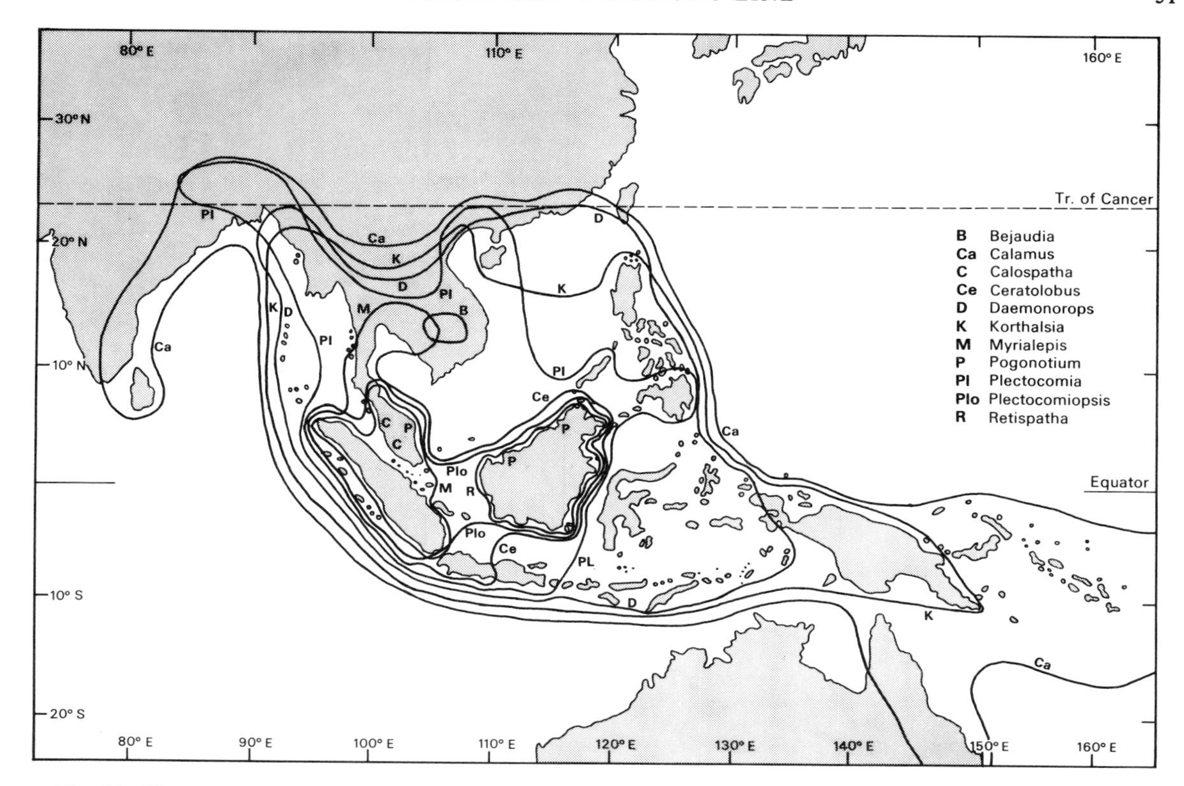

Fig. 6.7. The ranges of the eleven climbing palm (rattan) genera which occur in Malesia. *Calamus* also occurs in Africa.

perhaps, relatively recent speciation, which is probably still going on (as suggested by some extraordinary complexes of closely related taxa such as the complex of *C. perakensis* in Malaya and *C. javensis* throughout Sundaland). Unfortunately we still know too little to subdivide the genus satisfactorily, though this has been attempted by Beccari (1908) and Furtado (1956). The groupings of Beccari show more phytogeographic integrity than do the sections of Furtado and few generalizations can be made other than that most of the groups are found in Sundaland with few elsewhere. It seems that the radiation of *Calamus* in New Guinea and elsewhere east of Wallace's line is more likely to be due to eastwards migration of this very plastic genus after the Miocene collision.

There are, however, other Lepidocaryoid palms centred in Papuasia for which an austral path from old Gondwanaland seems the most reasonable origin. These are *Metroxylon* (Fig. 6.8) and *Pigafetta* (Fig. 6.1). *Pigafetta* has already been mentioned as one of the two Papuasian genera to reach Celebes. Although Moore (1973*b*) includes *Pigafetta* in the *Salacca* alliance, I believe its affinities to be with *Metroxylon* despite its dioecy. *Metroxylon* comprises about six species—two species in Fiji and Samoa, one in Carolines, one in New Hebrides, one in Solomon Islands, and one (*M. sagu*) native to New Guinea and the Moluccas but widespread as a cultivated palm. Its diversity east of Wallace's line is suggestive of an austral origin for the genus. There are, however, three points worth noting in relation to *Metroxylon*. The first is that we still know relatively little of the relationships of the genera within the Lepidocaryoid Major Group: the phytogeography certainly suggests that the relationships of *Metroxylon* with *Lepidocaryum*, *Mauritia*, and *Raphia* should be examined closely. The second is that *Eleiodoxa*, a monotypic genus formerly included in *Salacca* and since Burret moved it to a new genus (Burret 1942) now thought to be quite distinct, has many features suggestive of a close relationship with *Metroxylon*. *Eleiodoxa* is a palm of the Sundaland swamp forests. Is it as postulated for *Cyrtostachys*, a late or post Miocene immigrant from Papuasia? Clearly generic relationships need investigating. The third is that Beccari (1918), Corner

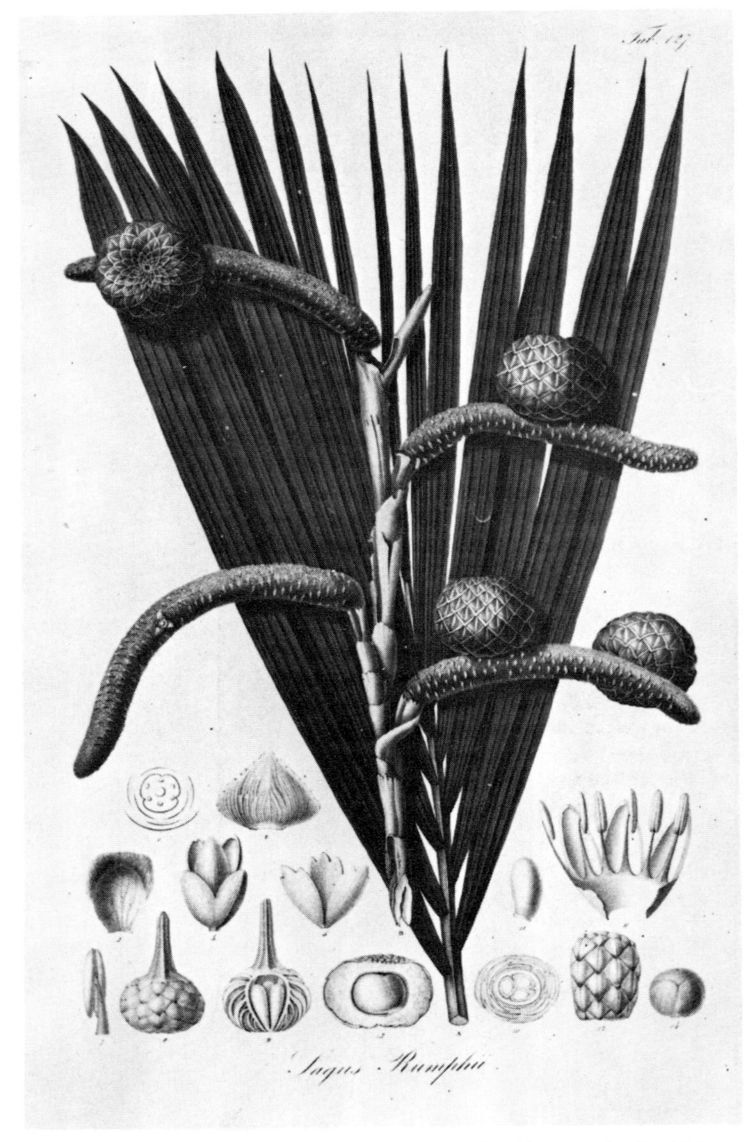

Fig. 6.8. *Metroxylon sagu*, the Sago Palm, a Lepidocaryoid palm. This genus is centred on Papuasia (from Rumphia 2, 1843, Table 127).

(1966), Whitmore (1977*a*), and Dransfield (1979) among others have all noted the similarity in inflorescence structure between the rattan *Korthalsia* (twenty-two species) and *Metroxylon*. *Korthalsia* is certainly quite different in many ways from the other rattan genera; its peculiarities suggest evolution of the climbing habit more than once. It is, however, concentrated on Sundaland, with two species only crossing Wallace's line. *Metroxylon* is, in nature, confined east of Wallace's line. Is the similarity in inflorescence structure between these two genera the result of parallelism or are they related? If related, have they reached Malesia by two different routes?

Eugeissona is a Lepidocaryoid genus of five non-climbing palms, confined to Borneo and the Malay Peninsula. There is a fossil record of two pollen-types in the genus from late Oligocene or early Miocene and middle Miocene in Borneo (Muller 1968, 1972); thus suggesting the existence of the genus in Laurasia before the collision with Gondwanaland. In view of this, any direct relationship with *Metroxylon* seems unlikely. Its ultimate origin in Malesia must be sought via a northern route.

Orania

The primitive Arecoid genus *Orania* is of interest in that it is present on both sides of Wallace's line. There are about sixteen species, eight of which are endemic to New Guinea, one species is known from north Australia, one species from Moluccas, and one or possibly two from Sundaland as far north as peninsular Thailand. The rest of the species are Philippine endemics (possibly reduceable to two species). The only record for Celebes is a photograph in Fairchild (1943) of an *Orania*, supposedly taken in Celebes. In isolation *Orania* is perhaps analogous to *Cyrtostachys* and *Eleidoxa*/*Metroxylon*—a genus of predominantly Papuasian distribution which migrated into Sundaland after the late Miocene collision. However, Moore (1973*b*) indicated a very close relationship between *Sindroa* of Madagascar and *Orania*, and this suggests a pattern of ultimate origin similar to the Clinostigmatoid alliance.

Oncosperma

Oncosperma, with one species crossing Wallace's line, and otherwise known from Ceylon, Sundaland, and Philippines, is most closely related to a group of palms including *Acanthophoenix*, *Deckenia*, *Nephrosperma*, *Phoenicophorium*, *Roscheria*, *Tectiphiala*, and *Verschaffeltia* of the Mascarenes (Moore 1973*b*, 1978). This distribution suggests a contrast to *Orania*, a pattern suggesting rafting northwards on the Indian Plate and thence migration eastwards to Malesia. *Oncosperma* fruits occur in the Eocene age London Clay (Chandler 1961–4) and it was already present as pollen in the Oligocene in Borneo (Muller 1972).

The forest coryphoid palms

The last genera to be considered in detail, *Licuala* and *Livistona* are perhaps the most problematical. They belong to the category of genera with major specific representation on either side of Wallace's line coupled

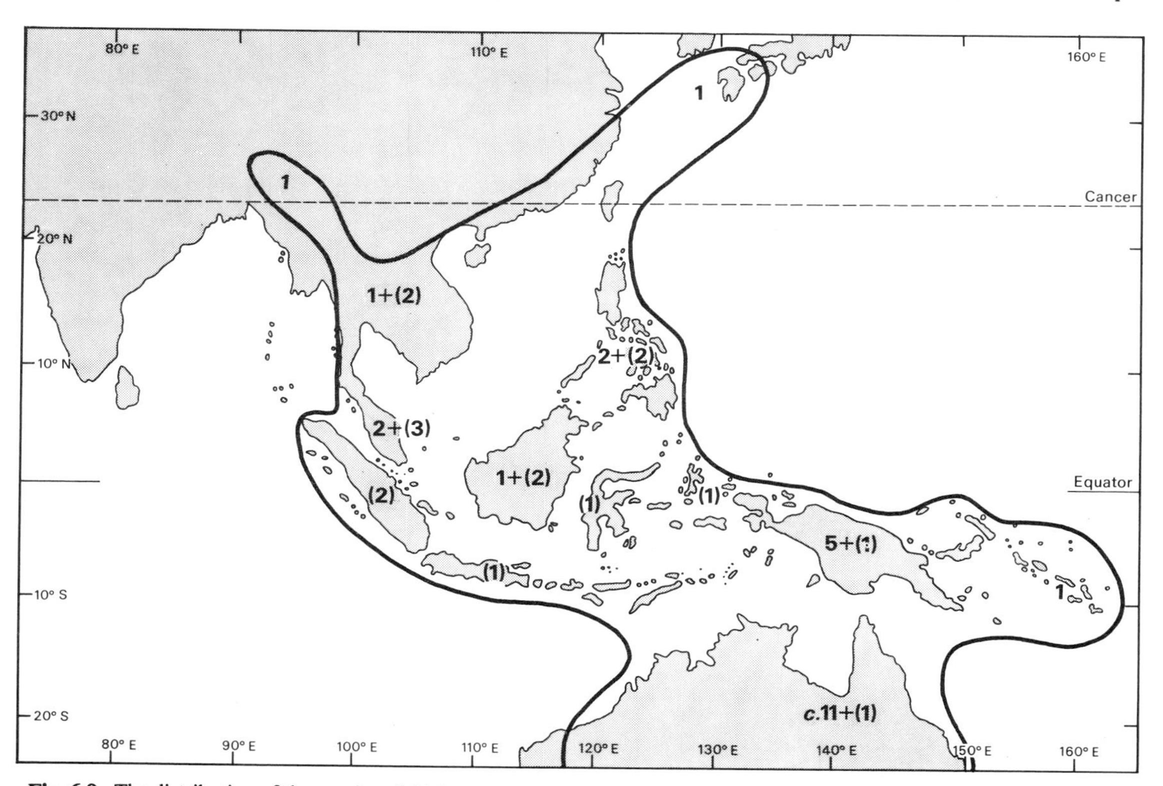

Fig. 6.9. The distribution of the species of *Livistona*. Numbers of endemic and, in parentheses, non-endemic species shown.

with very poor representation in Celebes. The two genera are closely related to each other and to *Pholidocarpus* in the Malesian region, and to other genera found in the Pacific (*Pritchardia*, *Pritchardiopsis*), the Americas (*Acoelorrhape*, *Brahea*, *Colpothrinax*, *Copernicia*, and *Serenoa*), and Arabia and the Horn of Africa (*Wissmannia*) (Moore 1973*a*). *Johannesteijsmannia* of Sundaland has flowers very similar to those of *Licuala* and *Livistona* but has been kept apart as a unit of its own because of differences in inflorescence structure, and *Washingtonia* of south-west USA and north-west Mexico has also been separated as a unit within the *Livistona* alliance by Moore (1973*a*). Whereas most members of the alliance are plants of semi-arid areas, *Johannesteijsmannia*, *Pholidocarpus*, and *Pritchardia* (and presumably the supposedly extinct *Pritchardiopsis*) and nearly all species of *Licuala* are plants of rain forest. *Livistona* has species adapted to semi-arid conditions as well as some rain forest species, including the diminutive *L. exigua*, an undergrowth palmlet of Brunei, Borneo (Dransfield 1977). Except for this last species *Livistona* are rather big shrub to tree palms. *Licuala* on the other hand consists mostly of undergrowth palmlets of rain forest. Most *Licuala* species are immediately distinguishable by their curious mixed reduplicate/induplicate leaves (Dransfield 1977; Whitmore 1977*a*). Apart from one or two exceptions, *Licuala* can be seen as the representative of the Coryphoid Major Group in the Malesian rain forest undergrowth, and *Livistona* as the representative in the forest canopy (along with *Pholidocarpus* in the western part of Malesia).

The paradox is that *Livistona* has almost equal representation at the two extremities of Malesia and *Licuala* also has an almost equal radiation (on a large scale) at either extremity, but with only one or two species in between (Table 6.1 f; Figs. 6.9, 6.10). In the instance of *Livistona* there is almost equal diversity in morphology as well as number of species in the two centres, and the phytogeography is further complicated by the presence of a scarcely distinct genus *Wissmannia* in the Horn of Africa and Arabia. Moore (1973*a*) regarded the Coryphoid Palms as being relatively primitive and that view is concurred with here. The present day distribution of *Livistona* (and closely related *Wissmannia*) may be

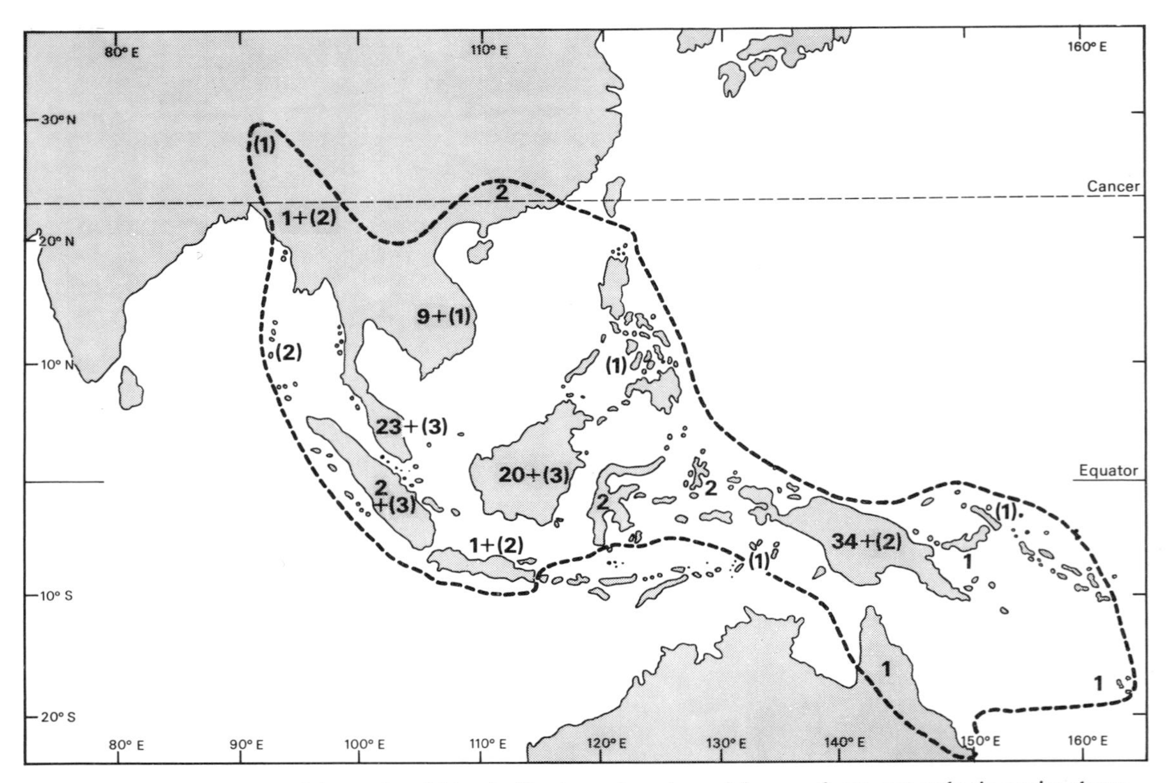

Fig. 6.10. The distribution of the species of *Licuala*. Numbers of species and, in parentheses, non-endemic species shown.

seen as a relict of a much wider distribution, which might have come about from a West Gondwanic origin arriving in Australia/New Guinea and Arabia/India/east Asia/Sundaland separately. The extraordinary relict distribution and great diversity in morphology in Australia seems incompatible with a post-Miocene immigration from Sundaland. Unfortunately, there are no fossils to help these rather wild speculations.

Licuala is perhaps still more difficult to explain. There are extraordinary radiations including local endemics in Sundaland (*c.* 50 spp) and New Guinea (36 spp), with further species in adjoining areas, yet only one or two species in Celebes. It seems very difficult to imagine this markedly bicentric distribution as being due to post Miocene, post collision, migration. If dispersal was west to east, why should there be such enormous diversity on each side of Celebes and Moluccas, but so few species in the middle? Could *Licuala* have invaded from both ends, or must climatic vicissitudes during the Pleistocene be used to account for extinction in Celebes? Perhaps questions of this nature should await a critical morphological and phylogenetic analysis of the genus. Of course, much of the above discussion could be made redundant, if genera were able to be dispersed between Laurasia and Gondwana before the collision but even then the paucity of species in the middle of the archipelago remains enigmatic. It is, however, the reasonableness of the explanation of the patterns of distribution in Malesia in terms of the Miocene collision of the plates, which makes the presence of long distance pre-Miocene dispersal so unlikely.

CONCLUSION

It can be seen quite clearly that the palm floras of west

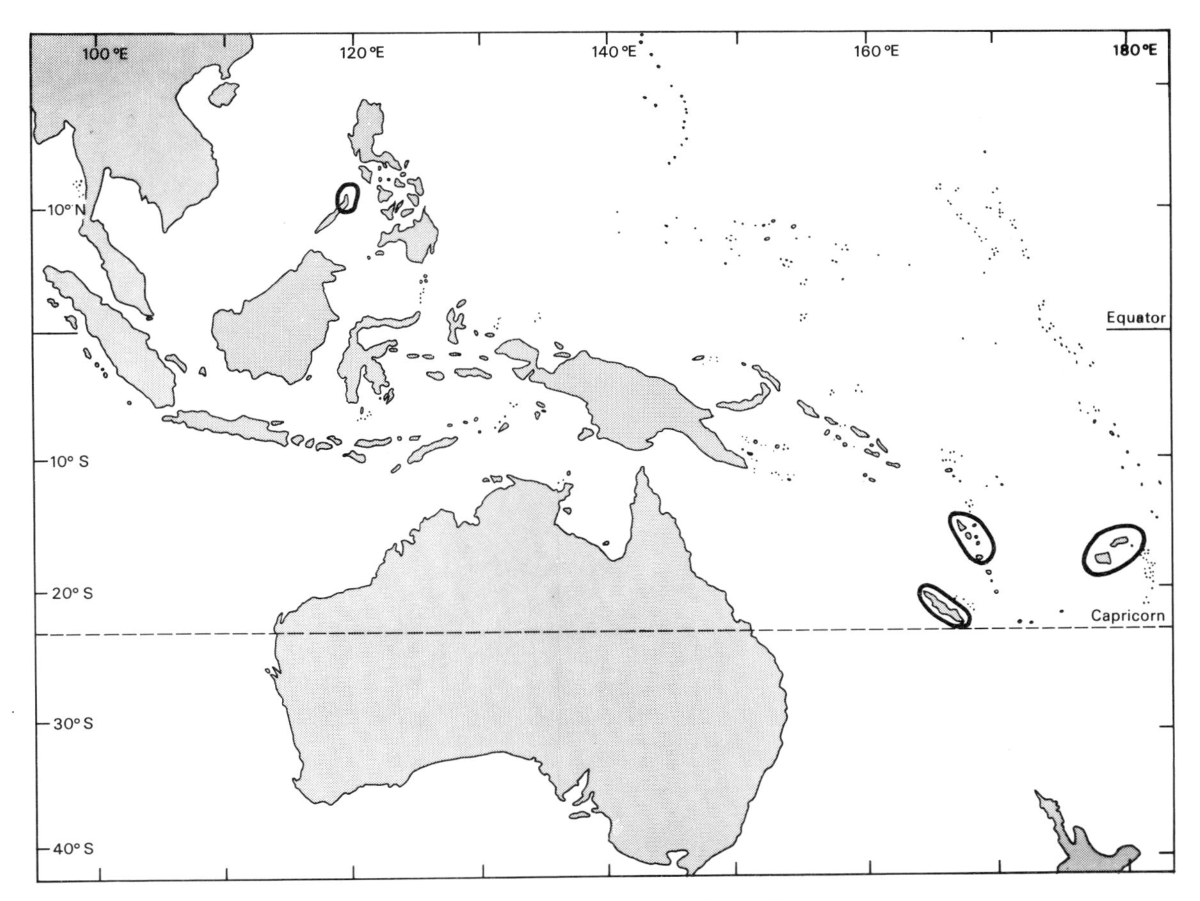

Fig. 6.11. The enigmatic range of *Veitchia*.

and east Malesia are in many ways markedly different, and that such differences would not occur if extensive dispersal across Wallace's line had taken place after the collision of the two plates in the Miocene. Some genera and even species have become widespread but their dispersal seems quite reasonably attributable to post-Miocene migration either in an easterly or westerly direction. In the rare cases where a fossil record exists, it does not contradict the hypotheses of post-Miocene dispersal. There are, however, some extraordinary patterns at the levels of Major Group, alliance or genus suggesting invasion of Malesia by northern and by austral routes after the breakup of Gondwana, an ancient origin in Gondwana being implicit.

There remain some genera the distribution of which seems at present quite inexplicable in terms of the gradual evolution of Malesia; two such are *Areca* and *Veitchia*. *Areca* is remarkably diverse in both Sundaland and Papuasia. The diversity in the two areas is quite different and there is some suggestion that the Philippine representatives of the genus might represent an intermediate between the two floras. Migration may have taken place via a route including the Philippines but we still know too little of relationships between the species. *Veitchia* is perhaps the most remarkably dispersed palm in the region (Fig. 6.11). The genus belongs to the *Ptychosperma* alliance (Moore 1973*b*) which is almost entirely Papuasian or western Pacific in its distribution. *Veitchia* has eighteen species one of which is only known from cultivation but which is probably from New Hebrides, along with four others; ten species are known from Fiji, one from New Caledonia, one of unknown origin but possibly from Australia, and one species, *Veitchia merrillii*, from the islands of Palawan and nearby Calamianes in the western Philippines (Moore 1957). This last is the only species in the genus and the whole alliance found west of Wallace's line. What is even more incomprehensible is that of all islands in the Philippines it is found on those nearest Sundaland and with the strongest Sundaic element in both plants and animals. As yet this strange distribution pattern cannot be accounted for. However, it is worth noting that other plants are known with largely east Malesian distribution and represented also in the Philippines, *Heterospathe* has already been mentioned in this context and *Sararanga* (Pandanaceae) also has this type of distribution.

Does Wallace's line exist for the Palm family? The answer must surely be, yes. The geological evidence of the Miocene creation of Malesia does much to explain the differences between the Papuasian and Sundaland palm floras. There are, however, many features of distribution not yet explicable, largely because we still know too little of intra- and inter-generic relationships, especially in the big genera such as *Calamus* and *Licuala*.

7 THE VERTEBRATE FAUNAS

The Earl of Cranbrook

Subsequent collecting has not invalidated Wallace's observations on the pronounced differences, in origin and composition, between the vertebrate faunas of Sundaland and Papuasia. In present times, all vertebrate classes are represented on the large islands of the Sunda Shelf by rich and distinctive assemblages of species of undoubted Asian affinity. Many species (and several higher taxa) do not extend further eastwards. Yet distributions within Sundaland, notably the progressive eastwards impoverishment within Java, indicate that in many cases ecological factors other than a sea barrier operate to limit dispersal. Moreover the terrestrial vertebrates of the Philippines, Celebes, and the Lesser Sunda Islands are predominantly Asian in affinity and, except for primary division freshwater fish, the eastern margin of the Sunda Shelf does not mark the boundary between Oriental and Australasian faunas. If any line is to be drawn, it should circumscribe Celebes, the extant vertebrate fauna of which shows many features characteristic of oceanic islands. The extinct fauna of Celebes suggests the existence of past immigration routes from Asia that are no longer available. For instance, the mammals included three proboscidean species, also present in deposits of Plio-Pleistocene age in Java, of which two (both stegodonts) were shared with Flores and Timor (with geographical representatives in the Philippines). The present world distributions of species of bee-eaters (Meropidae), kingfishers (Alcedinidae), and true rats (Rattus) *suggest that these groups originated in Malesia, perhaps soon after the region was formed by the collision of Laurasian and Gondwanan plates. Although Celebes lies near the middle of this region, there are no clear grounds to identify the island in its present shape and form as the centre of evolution of any one of these taxa.*

In his classical exposition of 1863, A. R. Wallace set the demarcation between the Oriental and Australasian faunas at the straits separating Bali from Lombok and Borneo from Celebes. Although in later life he revised his views (see Chapter 2) this remains the line associated with his name.

Wallace's evidence was drawn from observations of the known distribution of extant organisms, chiefly those which attracted his own attention, i.e. mammals, birds, and certain of the brighter insects. Possibly he was unduly influenced in his judgement by the occurrences of a few especially conspicuous members of these classes. Indeed, passages in *The Malay archipelago* testify to the strong impression made, soon after his arrival on Lombok (the eastern shore of his future line), by his first encounters with Lesser Sulphur-crested cockatoos, *Cacatua sulphurea*, and with certain other brightly coloured or behaviourally prominent members of avian groups characteristic of the Australasian region:

> Small white cockatoos were abundant, and their loud screams, conspicuous white colour, and pretty yellow crests, rendered then a very important features in the landscape. . . . Some small honeysuckers of the genus *Ptilotis*, and the strange mound-maker (*Megapodius gouldii*), are also here first met with on the traveller's journey eastward. . . . Here also I met with the pretty Australian Bee-eater (*Merops ornatus*).
>
> *The Malay archipelago* (1869, Vol. 1, pp. 243–5).

On Celebes, Wallace travelled widely and himself 'assiduously collected birds' (Wallace 1869, Vol. 1, p. 426). Combining his own collections with those of others gave a known total of 191 species from Celebes and associated small islands, of which 144 were land birds (128 from Celebes itself alone). Eighty land bird species he considered endemic (Wallace 1893). Of mammals he knew only seven bats and fourteen terrestrial species, which he listed as follows: two 'Eastern opossums', the 'curious Lemur' (Fig. 7.1), a 'curious baboon-like monkey', the 'common Malay Civet', five distinctive squirrels, the wild pig which 'seems to be of a species peculiar to the island', the Babirusa or Pig-deer' (Fig. 7.2), a deer 'which seems to be the same as the *Rusa hippelaphus* of Java, and was probably introduced by man' and the 'wild cow' (Fig. 7.3) (Wallace 1869, Vol. 1, pp. 432–6).

Later collecting and research have added to these lists. Knowledge has also been acquired on the composition and affinities of reptile, amphibian, and freshwater fish faunas, vertebrate groups not treated in detail by Wallace in his analyses. Drawing on information available up to the early 1950s covering all vertebrate classes, Darlington (1957), in an important synthesis, included a balanced appraisal of the zoogeographical significance of Wallace's line and its suggested alternatives at the border of the Oriental region with the Australasian.

Darlington's interpretations were consistent with the

Fig. 7.1. *Tarsius spectrum*, the Celebes endemic 'curious lemur' of Wallace.

Fig. 7.3. The larger of the endemic Celebes' wild cattle or anoas, *Anoa depressicornis*.

Fig. 7.2. The remarkable babirusa or pig deer, *Babyrousa babyrussa*. For a recent review, see Groves (1980*a*).

belief that the positions of the continents were essentially immutable. The subsequent acceptance of modern tectonic theory inevitably led to a reappraisal of some of the assumptions of biogeography. Models based on vicariance have been proposed (see, for example, Nelson 1974; Rosen 1975) and criticized (McDowall 1978).

New knowledge of the distribution and taxonomy of extant vertebrate species has done little more than alter details of Darlington's (1957) treatment. In palaeontology, however, excavations of recent decades have shown that there existed in Celebes and the Lesser Sunda Islands, in the Pleistocene and early Holocene periods, alongside the precursors of modern species, certain other unrelated vertebrates which have left no surviving descendants. The implications of the fossil evidence have been discussed in zoogeographical terms in relation to mammals by Hooijer (1975) and Groves (1976), and to some extent have been reviewed in the light of plate tectonic data by Audley-Charles and Hooijer (1973). It is useful to take further stock of the data in the fuller context of Chapter 4. Suggestions that certain widespread avian families, i.e., bee-eaters, Meropidae (Fry 1969), and kingfishers, Alcedinidae (Fry 1980*a*, *b*), or the cosmopolitan mammalian genus, *Rattus* (proposed *en passant* by Groves 1976), may have undergone initial radiation in the Malesian region, or even on Celebes itself, also demand evaluation in the light of the new palaeogeographical understanding.

It is equally important to give proper weight to ecological factors, past and present, in determining distributions that prevail today. Vertebrates exhibit great adaptive diversity, so that no single environmental factor is likely to be uniformly significant in its effects on dispersal for all species. It is clear that during the Quaternary period this region experienced considerable climatic vicissitudes, reviewed in Chapter 5. While not as dramatic as the ice ages of higher latitudes, the variations involved have been considered sufficient to account for the extinction of a large proportion of the middle Pleistocene mammalian fauna (Medway 1972, 1979).

The extent to which present distributions have been influenced by comparatively recent climatic changes also needs to be considered.

EXTANT VERTEBRATE FAUNAS

Freshwater fishes

There are obviously very strong ecological constraints on the dispersal of fishes of the freshwater environment. Those that are intolerant of salinity are closely tied to the waters of a continuous landmass, or even to a single catchment area. The distribution of obligatory freshwater fishes has for a long time been recognized as highly significant to zoogeography (Myers 1949; Bănărescu 1975).

It is customary to class the fishes occurring in freshwater into ecological groups defined by their tolerance to salinity. The simplest classification recognizes a primary division containing those that are strictly confined to freshwater, a secondary division, those that live chiefly in freshwater but show a little tolerance to salt, and a peripheral division containing those of considerable salt tolerance including, for instance, diadromous fish and fish of marine origin that have colonized the freshwater environment on oceanic islands (Darlington 1957). Although such classification is ecological and is based on empirical evidence, there are correlations with natural systematic groups. Thus, all members of the primary division are bony fishes, Osteichthyes. Those occurring in the Oriental and/or Australasian regions comprise members of the following groups: Dipnoi, represented by *Neoceratodus forsteri*, the one living species of the lung-fish family Ceratodontidae, now localized in Australia but more or less cosmopolitan in the Mesozoic; Osteoglossomorpha, represented by one genus of bony-tongues, Osteoglossidae, and one of feather-backs, Notopteridae; Ostariophysi, the dominant order of freshwater fish world-wide, represented here by many examples, notably among the cyprinoids (carps and allies) and siluroids (catfish); Acanthopterygii, i.e. leaf-fishes, Nandidae, climbing perches, Anabantidae, snakeheads, Channidae and spiny eels, Mastacembelidae. The most important members of the secondary division in this area are Atherinomorpha of the suborder Cyprinodontoidei, chiefly Cyprinodontidae.

It is well known that the freshwater fish faunas of Sundaland and Papuasia differ strikingly in composition. To the west of Wallace's line, the Greater Sunda Islands support a rich assemblage of primary division groups, notably Ostariophysi, Anabantidae, Channidae, and Mastacembelidae, as well as many secondary division cyprinodont species. Despite a considerable degree of local radiation, this fauna is clearly related to that of adjoining continental Asia.

The distributions of selected representative groups of primary division fishes illustrate the variety of patterns seen in the Sunda region. The examples given in Table 7.1 comprise the following: two cyprinid genera, *Chela* being widespread at tropical latitudes in continental Asia with one species reaching Malaya and Sumatra and one endemic on Sumatra, and *Rasbora* showing an important recent radiation centred in the area of southern Malaya

Table 7.1

The occurrence in Sundaland of selected groups of primary division freshwater fishes. (From Silas 1953, 1958, Brittan 1954, Sufi 1956, and Inger and Chin 1962.) Figures in the body of the table denote numbers of species known from each land area, with (in parentheses) the number of local endemics.

Taxon	*Total species*	*Malaya**	*Sumatra*	*Borneo*	*Java*
Cyprinidae, *Chela*	2	1(0)	2(1)	0	0
Cyprinidae, *Rasbora* (Fig. 7.1)	34	18(1)	22(4)	18(8)	3(1)
Homalopterinae	21	7(3)	13(8)	7(3)	5(0)
Gastromyzontinae	8	0	0	8(8)	0
Mastacembelidae (Fig. 7.2)	9	7(2)	5(0)	5(1)	3(0)
TOTALS	74	33(6)	44(13)	38(20)	11(1)

* i.e. Peninsular Malaysia, Singapore and peninsular Thailand south of 10°N.

and the east coast of Sumatra (Fig. 7.4); the hill stream loaches, Homalopterinae, with only one continental species extending beyond the Malay Peninsula, but an important local radiation producing five species more or less widespread in Sundaland and a total of 14 endemic to single landmasses; the sucker loaches, Gastromyzontinae, which have their continental centre of radiation in China and occur in Sundaland only on Borneo as a diverse assemblage of endemic species; and finally the spiny eels, Mastacembelidae (Fig. 7.5). The last-named combine all patterns, in a manner characteristic of many vertebrate groups of the region, showing continental connections most strongly in the Malay Peninsula and Sumatra, and clear evidence of a Sundaic radiation with some regional endemics more or less widespread in Sundaland and others confined to single landmasses. Both the spiny eels and the general totals of the selected fish taxa (Table 7.1) illustrate the comparative poverty of the fauna of Java, which consists chiefly of widespread Oriental or Sundaic species and is deficient in endemics.

To the east, New Guinea freshwaters also support a rich fish fauna comprising some 207 species of which about 111 are permanently resident in this habitat. Yet only one, the bony-tongue *Scleropages jardini*, is placed in the primary division and there are no secondary division representatives; all others are species of the peripheral division. The New Guinea fauna is closely related to that of Australia, which includes *Scleropages* and a second primary division genus, the lung-fish *Neoceratodus forsteri* (see above), but again lacks secondary division freshwater fishes.

Intermediate in location, the freshwater fish fauna of Celebes also consists very largely of examples of the peripheral division. Some secondary division species of Asian origin are present (i.e. the cyprinodont genus *Aplocheilus*) and cyprinodont stock has radiated locally to form the endemic family Adrianichthyidae (Darlington 1957). In addition, the snakehead *Ophicephalus striatus* and the climbing perch *Anabas testudineus* are found in Celebes (and also extend along the Lesser Sundas to Flores and to the Moluccas). These fishes, however, are equipped with accessory air-breathing organs and—partly as a consequence of this adaptation—are food-fish of some importance in the region, often carried alive from place to place. Zoogeographers have concluded that they are likely to have been spread artificially (cf. Darlington 1957). There is thus no reliable record of a spontaneously occurring primary division freshwater fish in Celebes. The Makassar Strait appears to have acted as an effective barrier to eastward movement of the rich

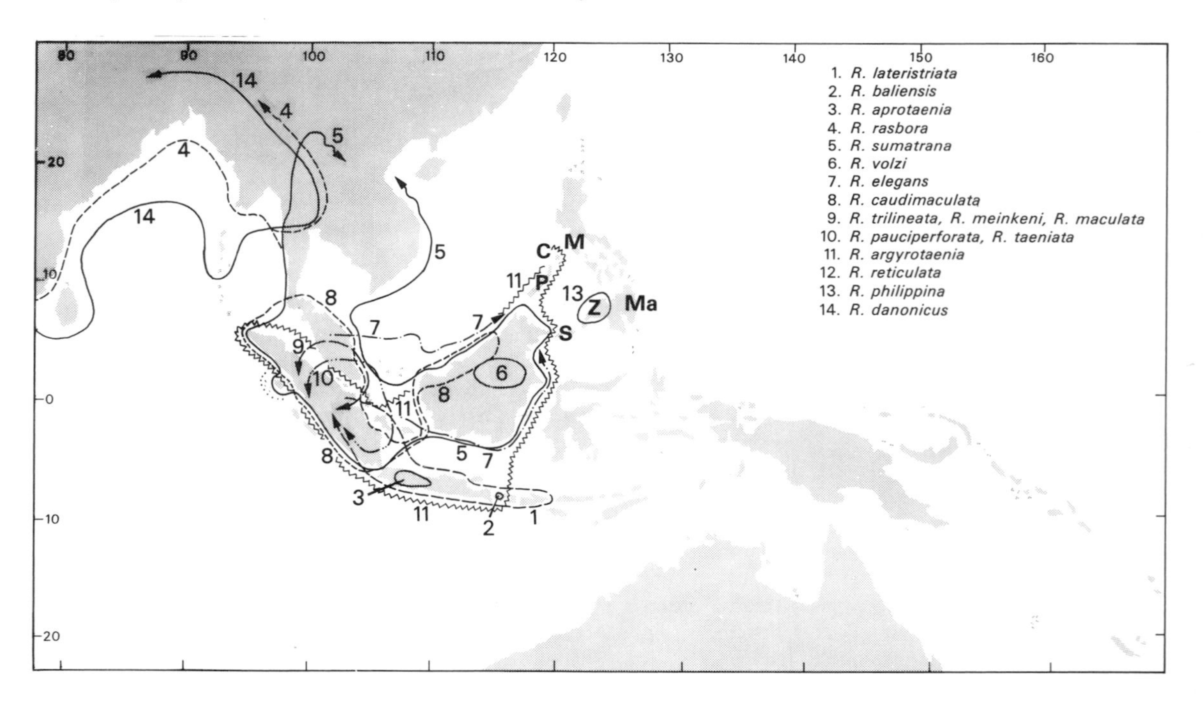

Fig. 7.4. Generalized distributions of selected species of the freshwater fish genus *Rasbora*, in the Oriental region. (From Brittan 1954.) **C**: Calamaian islands; **M**: Mindoro; **Ma**: Mindanao; **P**: Palawan; **S**: Sulu islands; **Z**: Zamboanga peninsula.

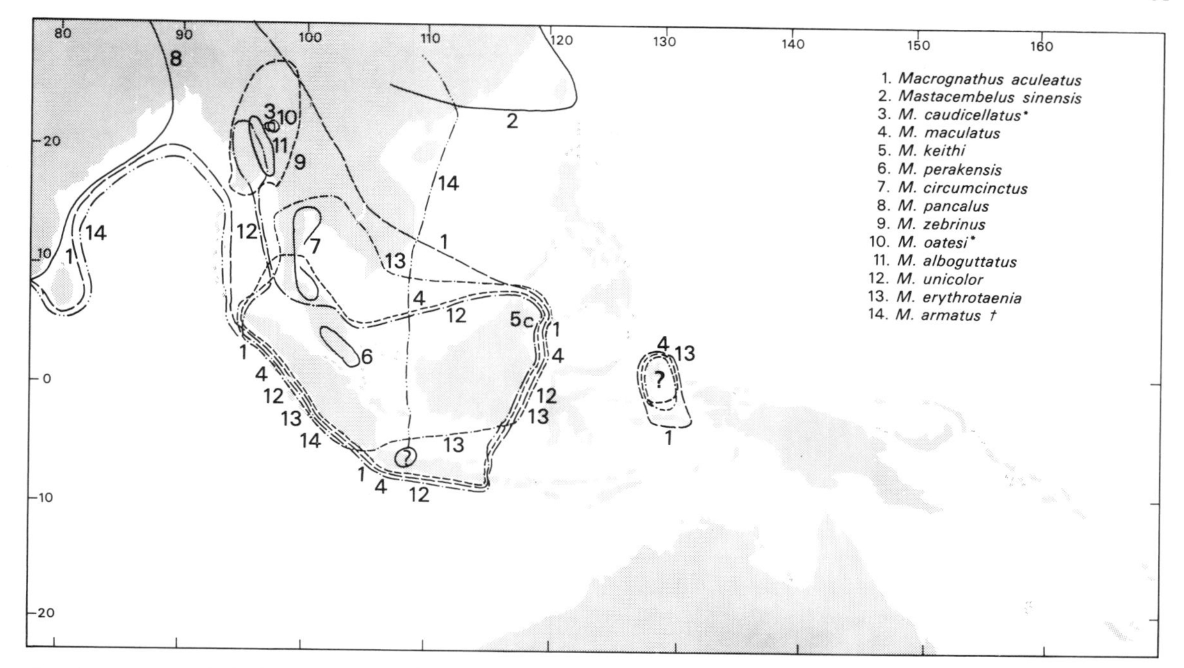

Fig. 7.5. Generalized distributions of the spiny eels Mastacembelidae in the Oriental region. From Sufi (1956). * Two endemic species occur in Lake Inle, Burma. † The authenticity of the records of *M. armatus* from Java was questioned by de Beaufort (in Weber and Beaufort 1911, Vol. 11).

assemblage of Borneo, and to this extent Wallace's line sharply demarcates the limits of an Asian fauna.

To the north and the south of Celebes, however, there has been limited eastwards migration of Sundaic primary freshwater fish. A few primary division Ostariophysi have invaded the Philippines, apparently by two routes, i.e. along the line Palawan–Mindoro and via the Sulu Islands to Mindanao. For example, among *Rasbora* (Fig. 7.4), the pan-Sundaic *R. argyrotaenia* is present along the axis from Palawan to Culion and Busuanga in the Calamaian islands, *R. semilineata* occurs only in northern Borneo and Palawan while *R. philippina*, an isolated endemic (but related to *R. argyrotaenia*), is confined to the Zamboanga peninsula of Mindanao (Brittan 1954). The distinctness of these two routes is also supported by the different representation of catfish: an endemic silurid in Palawan and the Calamaian islands and an endemic clariid on Mindanao (Darlington 1957).

In the south, similarly, small numbers of primary division Ostariophysi occur east of Wallace's line although there has been no local radiation. The cyprinid genera *Puntius* and *Rasbora* are each represented by one species on Lombok, and the *Rasbora* extends to Sumbawa. The catfish *Clarias batrachus* also extends along the Lesser Sunda chain. This fish, however, like the snakehead and climbing perch mentioned above, possesses an accessory respiratory organ and is thereby able to survive long periods out of water. It is a common food-fish and can be considered likely to have been carried by man.

Although the number of species involved is small, there is thus progressive attenuation of the primary freshwater fish faunas of these islands, which contrasts with the abrupt termination at the Makassar Strait. At the Lombok Strait there is a reduction in the fauna, but the final limit of naturally occurring primary division freshwater fish is ultimately reached at Sape Strait east of Sumbawa. The process of subtraction, however, begins west of Wallace's line. The representative groups tabulated and figured illustrate the impoverishment of Java by comparison with Borneo or Sumatra. There is a further reduction in total faunas from Java to Bali. Yet there is also evidence of attenuation of the fish fauna within the island of Java, from west to east. For instance, of the *Rasbora* species, the endemic *R. aprotaenia* is known only from western Java (Brittan 1954). The *Rasbora*-fauna of Bali is in fact richer than that of the adjoining eastern parts of Java (Fig. 7.4).

Amphibians (Fig. 7.6)

In contrast to the freshwater fishes, non-marine members of all other vertebrate classes have succeeded in crossing Wallace's line to reach Celebes. Most closely tied to the freshwater habitat are the amphibians. The known fauna of Celebes is not large: twenty-three species of anurans

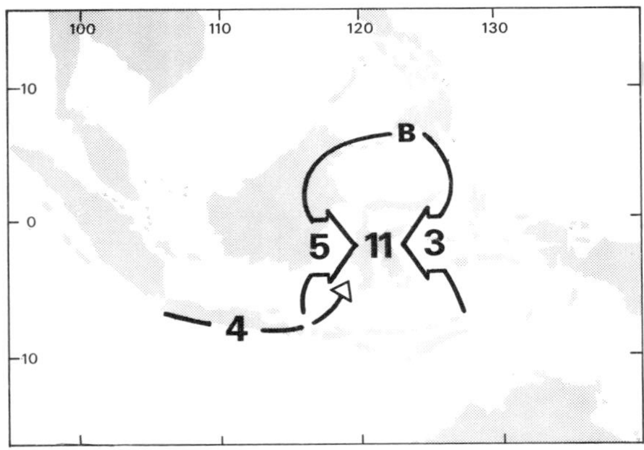

Fig. 7.6. Affinities of the amphibian fauna of Celebes. **B**: Basilan.

(frogs and toads), of which two are in doubt (Table 7.2). This figure may be compared with 100 species known from Borneo, including five caecilians—a group which does not occur on Celebes (Taylor 1968). Of the eight frogs that are common to Celebes and Borneo, two are questionable records and the remainder widely distributed and ordinarily associated with man-made habitats. Potential assisted dispersal in such cases is illustrated by a specimen of *Kaloula pulchra* taken aboard a ship carrying agricultural produce (Inger 1966). Inger judged it no coincidence that the only authentic records of this species in Borneo and in Celebes were from ports. *Rana erythraea*, the common paddy-field frog of the Sunda region, has been taken on Celebes only in the neighbourhood of Makassar, the centre of the most densely populated rice-growing area of the island (Inger 1966, and personal communication). Both frogs may be very recent introductions to Celebes. In effect, the ecology of the species held in common emphasizes the extent to which the Makassar Strait marks the natural limits of an Asian fauna.

Of the remaining amphibians of Celebes, twelve, (i.e. 57 per cent of the authenticated total) are endemic to the island. Of these, four show clear Sundaic affinities and two Papuasian. Finally, three species extend eastwards to islands of the Moluccas or further. These are listed as Papuasian in Table 7.2; the origin of two may in fact have been on Celebes and their present distribution the result of subsequent dispersal. Thus, discounting apparently recent (and continuing) assisted invasions of 'commensal' frogs, the Celebes amphibian fauna is poor in relation to island area (180 000 km²), and largely autochthonous at species level. Where affinity is identifiable, western origins predominate but eastern elements are also present.

As among fishes, there is no such abrupt faunal transition to the north and south of Celebes. Three caecilian

Table 7.2

Amphibian fauna of Celebes. Compiled from an annotated list prepared by R. F. Inger based chiefly on van Kampen (1923), Smith (1927), and Inger (1966 and unpublished). Species marked with an asterisk (*) also occur on Borneo

Family	*Sundaic species*	*Endemic species*	*Papuasian species*
Bufonidae	*(*Bufo biporcatus*)[1]	*Bufo celebensis*	
Microhylidae	**Kaloula baleata*[2]	*Oreophryne variabilis*	
	**K. pulchra*[2]	*O. celebensis*	
Ranidae	**Rana cancrivora*[2]	*Rana* sp. *microdisca* gp.	*Rana papua*
	**R. kuhli*[3]	*R. arathooni*	*R. grunniens*
	**R. erythraea*[2]	*R. microtympanum*	*R. modesta*
	**R. chalconota*[2]	*R. celebensis*	
		Ooeidozyga semipalmata	
		O. celebensis	
Rhacophoridae	**Polypedetes leucomystax*[2]	*Rhacophorus georgi*	
		R. edentulus	
		R. monticola	

[1] Doubtful; a single record, probably misidentified *Bufo celebensis* (R. F. Inger, personal communication.)

[2] These species frequent cultivation, disturbed or degraded habitat and are more or less closely associated with man. They are likely to have been assisted in dispersal by man, in some cases very recently; see text.

[3] The identification is questioned by Dr Inger, who suggests that the specimens may in fact be *R. microtympanum.*

species are known from the Philippines, two of *Ichthyophis* in Mindanao (one of which is also in Basilan) and one of *Caudacaecilia* in Palawan. Both these genera are present in northern Borneo but separate independent colonization of the Philippines is indicated. Twenty-two species of anurans known in Borneo also occur in the Philippine islands (Inger 1966). This total includes *Bufo biporcatus, Kaloula baleata, Rana cancrivora, R. erythraea*, and *Polypedetes leucomystax* (see Table 7.2) together with other species commonly associated with man-made habitat. Nevertheless, in many cases unassisted dispersal has undoubtedly taken place.

Along the Lesser Sunda chain there is progressive depletion of the Sundaic fauna, and a compensatory enrichment with Papuasian species, in which the Lombok Strait does not mark a transition any more decisive than

Table 7.3

The known land snakes of Celebes. The list is compiled from de Haas (1950) modified by Inger and Marx (1965). The wholly aquatic groups Hydrophiinae and Homalopsinae have been excluded. Species marked with an asterisk (*) were recorded in Borneo by Haile (1958)

Family	*Widespread Indo-Australian*	*Oriental*	*Celebes endemics or near-endemics*	*Papuasian*
Typhlopidae	*_Ramphotyphlops braminus_[1]		_Typhlops conradi_	_Typhlops ater_
Aniliidae		*_Cylindrophis rufus_	_C. celebensis_	
Boidae		*_Python reticulatus_		_Liasis boa_
		*_P. molurus_		_Candoia carinata_
		*_Xenopeltis unicolor_		
Acrochordidae	*_Acrochordus granulatus_			
Colubridae		*_Natrix trianguligera_	_Amphiesma celebica_	
		*_Rhabdophis chrysargus_	_A. sarasinorum_	
		*_R. chrysargoides_		
		*_R. subminutus_		
		*_Xenochrophis piscator_		
		*_X. vittatus_		
		*_Macropisthodon rhodomelas_		
		*_Elaphe flavolineata_	_E. erythrura_[2]	
		*_Gonysoma oxycephalum_	_G. janseni_	
		*_Chrysopelea paradisi_	_C. rhodopleuron_[3]	
		Ahaetulla prasina	_Ptyas dipsas_[4]	
		*_Dendrelaphis caudolineatus_		
		*_D. pictus_		
		Lycodon aulicus	_L. stormi_	
		*_Oligodon octolineatus_	_O. waandersi_	
		Calamaria virgulata	_Calamaria_ 7 spp.	
		*_Pseudorabdion longiceps_	_P. sarasinorum_	
			Rabdion forsteri	
			Calamorhabdium acuticeps	
		*_Boiga dendrophila_	_B. flavescens_	_B. irregularis_
		*_B. multimaculata_		
		*_Psammodynastes pulverulentus_		
Elapidae		_Bungarus candidus_		
		*_Naja naja_		
		*_Ophiophagus hannah_		
		*_Maticoa intestinalis_		
Viperidae		*_Trimeresurus wagleri_		

[1] Probably Oriental in origin; its spread eastwards, as far as Fiji, has undoubtedly been assisted by man.
[2] Also occurring in the Philippines.
[3] Ranging east to Banda and the Aru islands.
[4] Ranging east to Halmahera.

that between other pairs of neighbouring islands. Indeed, as among primary division freshwater fishes, the reduction of the fauna begins well to the west of Wallace's line. Caecilians occur in Java, but have not been found in Bali (Taylor 1968). For anurans, the figures compiled by Inger (1966) show that the fauna of Java (thirty-six species) is impoverished by comparison with those of the other main landmasses of the Sunda subregion: Malaya to 10°N (80), Sumatra (59), and Borneo – now known to be at least 95 (J. C. Dring, personal communication). Within Java, the fauna of the western part of the island (33–4 species) is richer than that of the east (21). Ten members of the east Javan fauna are absent from Bali; a further five are absent from Lombok, although the Sundaic *Rana macrodon* occurs on this island (as also on Sumbawa and Flores) but apparently not on Bali. Seven Sundaic species reach Flores, the same number as on Lombok (although forming a slightly lower proportion of the total amphibian fauna as a result of further enrichment by Papuasian species). The progressively diminished representation of species of Sundaic affinity is shown in percentage terms in Table 7.4.

Table 7.4

Percentage representation of species of Sundaic origin on the islands of the Lesser Sunda island chain crossing Wallace's line. Data for amphibians from Inger (1966), reptiles from Mertens (1930), and birds from Mayr (1944)

	Bali	Lombok	Sumbawa	Flores
Amphibians	92	78	75	70
Reptiles	94	85	87	78
Birds	87	73	68	63

Reptiles

The principal features of these dispersal patterns recur among higher vertebrates, modified in relation to the facility with which they cross the sea. Thus, for example, the number of snakes recorded from Celebes is only 61, compared with Peninsular Malaysia 136 (Grandison 1968), Sumatra 150, Java 110, or Borneo 166 (Haile 1958). The full list is shown in Table 7.3. It can be seen that the number of endemic species is comparatively high (17, i.e. 28 per cent), yet all but two endemics are members of genera also occurring in Sundaland. The non-endemic 'terrestrial' snakes (some of which are quite at home in the water) include two widespread Indo-Australian species and four clearly Papuasian species, but are otherwise chiefly widespread Oriental or Sundaic in distribution. Of the Oriental assemblage, most (20 species) are not recorded east of Celebes; the remainder (9) extend to one or more islands of the Moluccan region. Borneo and Celebes share 29 species of snake, 25 being terrestrial types. Thus, although it must not be assumed that immigrants necessarily crossed the Makassar Strait, 17 per cent of the Borneo snake fauna (or 46 per cent of the Celebes fauna) can be found on both sides of this sector of Wallace's line. The Lombok Strait presents an even less significant barrier. Referring to Lombok, Barbour (1912, p. 31) wrote: 'So far as the reptiles are concerned, the island is faunistically as Malayan as Bali' (see Table 7.4).

Birds

The bird fauna of Celebes has been much discussed in terms of its zoogeography. Stresemann (1939) identified 220 breeding species, of which 84 (38 per cent) he considered endemic; there has been no subsequent review of the island's avifauna. Comparable figures for Borneo are 396 species and 28 (7 per cent) endemics (Smythies 1968). While the richer fauna of Borneo is more closely allied to those of other parts of Sundaland, the Oriental affinity of the birds of Celebes is recognized (Mayr 1944). Some resident species, representing a diverse group of families, occur both sides of Wallace's line. In the south, once again, the process of depletion of the Sundaic fauna commences within the island of Java from west to east and continue stepwise along the Lesser Sunda chain. The Lombok Strait marks the limits of distribution of a comparatively large number of species but as far east as Flores species of Oriental affinity form the majority of birds. The process of faunal transition was discussed in detail by Mayr (1944); his diagrams and tables have often been reproduced (cf. Table 7.4).

Mammals

For the mammals of Celebes uncertainties over taxonomic status, particularly among the bats and the rats (see, for example, comments by Musser 1971), obscure faunal relations. The tabulations of Groves (1976) provide recent perspective. About 108 species are now recognized, representing 7 orders (including 41 bats); for comparison, the fauna of Borneo consists of 200 species, representing 11 orders (including 71 bats). The Celebes fauna is very distinctive. Some 71 species (66 per cent, including 17 bats) are believed to be endemic and a further 9 (7 bats) are shared only with the Moluccas. Seventeen endemic genera or subgenera have been recognized, including four bat genera; 15 are monotypic.

The bat fauna includes several widespread Indo-Australian species. Among other non-endemic bats, those with Sundaic affinities outnumber the Papuasian, but the latter are well represented. Among the non-flying mammals, however, only the two endemic species of the marsupial genus *Phalanger* (the 'eastern opossums' of Wallace) are of Papuasian affinity; all others, including endemics, are clearly of Asian derivation. The 14 non-endemic, non-flying, mammals all occur also in Sundaland; 12 occur in Borneo. This group consists of common, widespread commensals of man (house shrew, house rat, house mouse), pests of cultivated land (rats, porcupine, and the squirrel *Callosciurus notatus*)[1] or species with traditions of use or domestication by man (*Callosciurus prevostii*,[1] civets and the Javan rusa, *Cervus timorensis*). Several of the last group of species have been considered by zoogeographers likely to have been introduced. At a higher taxonomic level, the genus *Tarsier* is shared with the Philippines, Borneo and Sumatra, and *Haeromys* with Borneo only; other non-endemic genera are more widespread. None the less, the Makassar Strait clearly marks the eastern limits of the distinctive Sundaic mammal fauna, shared by the Malay Peninsula, Sumatra, Java and Borneo. This fauna is exclusively Asian in origin, but characterized by local radiation among many families and genera, producing large numbers of regional endemics. The treeshrews, Tupaiidae, are one such family. The distributions of treeshrew species illustrated in Fig. 7.7 would be broadly matched by other families of small mammals (e.g. Sciuridae, Muridae, etc.). Within Celebes, there have been local radiations, involving in particular the genera *Macaca*, *Callosciurus*, and *Rattus*. The pattern of speciation among the macaques on Celebes has been explained in terms of periodic interruptions of gene flow, perhaps by fragmentation into several smaller islands (Fooden 1969; Groves 1980*b*). An alternative explanation could be based on the postulate of a limited number of invasions by representatives of the ancestral type, widely separated in time.

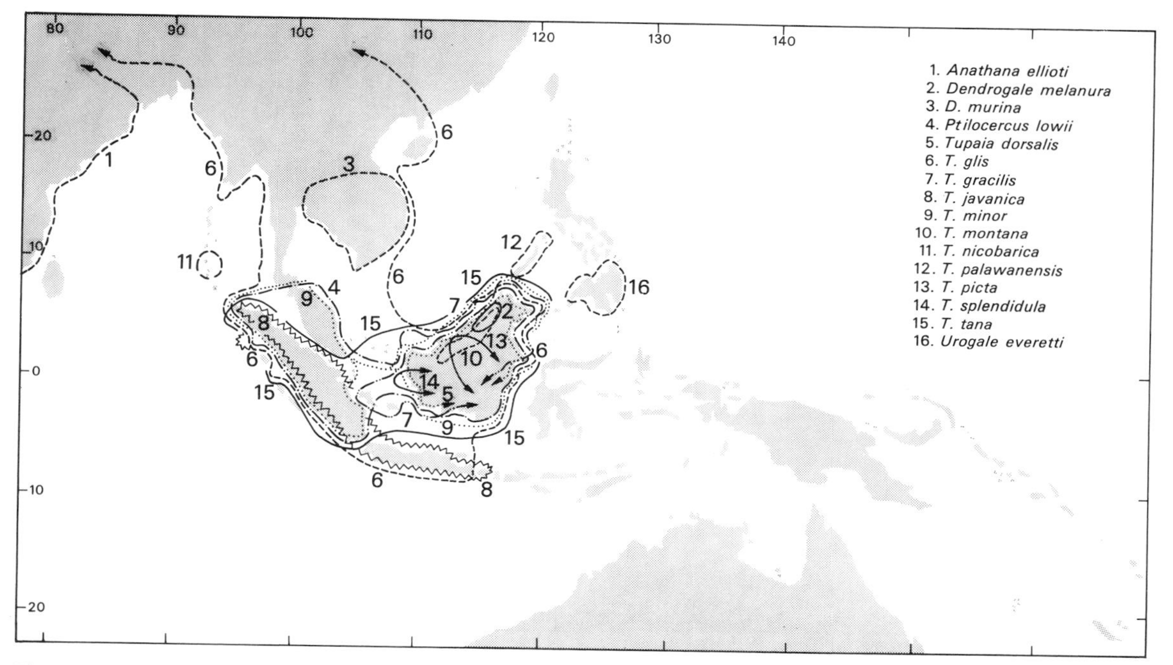

Fig. 7.7. Generalized distribution of treeshrews Tupaiidae. Chiefly from Lyon (1913), Medway (1977), Boonsong and McNeely (1978). Sundaic species (and endemics): Malaya 3(0), Sumatra 5(0), Borneo 10(5), Java 2(0).

The mammal faunas of the Lesser Sunda Islands are very poor. Sundaic species are present, but in many cases are thought to have been introduced in historic times (civets, deer, monkeys, etc.). The Philippines fauna is richer, including endemics; affinities clearly lie with Asian, particularly with Sundaic, mammals, and there is little evidence of significant interchange with Celebes (Groves 1976).

[1] Neither of these squirrels is widespread in Celebes; *C. notatus* is not recorded on the mainland (Laurie and Hill 1954).

EXTINCT FAUNA

Understanding of the extinct fauna of this region owes a great deal to the work of D. A. Hooijer (see bibliography). The principal collections from Celebes originated from Beru, Sompoh district, in the south-west. Results add three genera to the known mammal fauna: two proboscidean genera, *Stegodon* (Fig. 7.8, represented by a large form *S*. cf. *trigonocephalus* and a dwarfed form, *S. sompoensis*), and *Elephas* (represented by *E. celebensis*, another dwarfed species); and a unique, endemic, extinct giant pig *Celebochoerus heekereni* which shows no clear relationship with the Pleistocene pigs of Java (Hooijer 1954). In cave deposits at Bola Batu, remains of living species have been found, outside their present ranges: the endemic civet *Macrogalidia musschenbroekii*, the babirusa and larger anoa, *A. depressicornis* (Hooijer 1950, 1960): also known are fossils of the warted pig *Sus verrucosus* (Hooijer 1969*b*). Stratigraphic details do not give a clear age for these fossils. The proboscideans occur in Java in beds of Pliocene and Pleistocene ages (Hooijer 1969*a*, etc.). The Bola Batu cave deposits are probably Holocene (Hooijer 1950). Remains of an extinct giant land tortoise have also been found (Hooijer 1948), now identified as *Geochelone atlas*, which occurs elsewhere in Pleistocene deposits in the Siwaliks, northern India, in Java and Timor.

None of these fossil vertebrates has been found in Borneo. Their absence, however, need not be significant but may merely reflect the lack of appropriate conditions for preservation. Timor and Flores, on the other hand, have both yielded fossils of a large and a small stegodont, evidently the local representatives of those on Celebes. A similar pair of species is also known from the Philippines, including the islands of Mindanao and Luzon. Recently a dwarf *Stegodon* has been discovered in Sumba (Sartono 1979). There are also Pleistocene records of a large *Varanus* from Timor (and Java) comparable in size to the living Komodo 'dragon' *V. komodoensis* (Hooijer 1972). Cave deposits of late upper Pleistocene or Holocene age on Timor and Flores have yielded the remains of large rats, including extinct forms (Hooijer 1957*b*, 1965).

Fig. 7.8. Stegodonts, a hypothetical reconstruction. In the male the tusks are much longer than in the female and are adpressed for most of their length leaving no room for the trunk between them (Hooijer 1973, Fig. 5).

DISCUSSION

The peculiarity of the Celebes vertebrate fauna

It is indisputable that the vertebrate fauna of Celebes is anomalous in many respects. Although separated by no great distance from Borneo and Java, Celebes is not occupied by the distinctive Sundaic regional fauna that these islands share with the Malay Peninsula and Sumatra. In the generally impoverished selection of genera and families and the high incidence of endemicity, the Celebes vertebrates show many of the characteristics of oceanic island faunas (MacArthur and Wilson 1967; Gorman 1979).

The absence of primary division freshwater fish, other

than species likely to have been assisted in their dispersal by man, is a significant feature contributing to the 'oceanic' character. Cumulatively the evidence of all vertebrate groups strongly suggests that since its (most recent) emergence there has been no direct, unbroken subaerial connection between Celebes and the principal landmasses of the Sunda shelf. All immigrants from Sundaland have probably been obliged to cross a sea barrier. Notwithstanding geological evidence of Celebes' dual origin, its Laurasian segment has apparently not served as a raft carrying with it any strictly terrestrial (including freshwater) vertebrate of Asian origin. In their present composition, the vertebrate faunas provide no evidence to support the suggestion that the Makassar Strait was closed for a period during the late Pliocene (cf. Chapter 4). Moreover, any Quaternary land connection formed between Borneo and Celebes as a consequence of lowered sea-level appears likely to have been incomplete or short-lived. It was, in any event, apparently inadequate to facilitate the passage of primary division freshwater fish.

The Celebes faunas of all vertebrate classes are evidently derived from a limited number of colonizations spread over the period of subaerial exposure. The most ancient are now represented by endemic taxa and the most recent are the widespread 'tramps' (Diamond 1970) that so readily invade new habitat created by human activity. Because the sea barrier to be crossed is comparatively narrow (and may have been narrower at past times; Chapter 4) the closeness with which the Celebes fauna is related to the Sundaic varies with the dispersive abilities of the different taxonomic groups. Land birds in particular cross sea barriers with great facility (see, for example, the commentary by Lack 1976: Chapter 1). It is notable that for birds the ratio between the logarithms of number of species and island area for Celebes falls only a little below expectation when compared with Java, Sumatra, Borneo, and New Guinea (MacArthur and Wilson 1967, Fig. 9), whereas for mammals or amphibians it would undoubtedly be much lower.

The Makassar Strait thus demarcates the natural limit of a distinctive vertebrate fauna. It does not, however, separate an Asian fauna from an Australasian. As previous zoogeographers have recognized, and this review has reiterated, the fauna of Celebes is predominantly Asian in affinity. In all vertebrate classes except the fish of freshwater, species of Asian origin greatly outnumber those of Papuasian affinity. From the east, successful immigration of strictly terrestrial, non-flying vertebrates to Celebes has evidently been even less easy than from the west. The zoogeographical evidence thus suggests that it is very unlikely that the Gondwanan portion of Celebes was at any time since its final exposure linked by a continuous subaerial land bridge to landmasses further east. It follows from these distribution patterns that any zoogeographical 'line' to be drawn must in fact circumscribe Celebes (with adjacent, faunistically related small islands) to emphasize its peculiar isolation from the east as much as from the west.

Evidence of vicarious distributions

The presence of a lung-fish in Australia is explicable in terms of continental drift. The group is ancient and the precursors of *Neoceratodus* are widespread in the Triassic fossil record, dating from a period before the fragmentation of Gondwanaland. It is tempting to apply the same explanation to the presence of bony-tongues, *Scleropages* sp., in Australia and Papuasia. The Osteoglossidae too are primitive bony fishes whose few species are today confined to tropical regions of the globe. The one other living member of the genus, *S. formosus*, occurs on continental south-east Asia and in Sundaland. It is conceivable, among the osteoglossids, that species still sufficiently alike to be considered congeneric, have none the less been separated since late Cretaceous times and have independently ridden Laurasian and Gondwanan plates to arrive vicariously at their present locations of comparative proximity. On the other hand, this distribution could also have arisen by transgression of Wallace's line at some time in the late Tertiary or more recently. The direction of any such colonization cannot be determined with certainty. It may be significant that a preliminary trial of the salt tolerance of an adult *S. formosus*, reported by Roberts (1978), has demonstrated the capacity to survive in moderate salinity. The fish ceased to feed only when salt concentration reached 13–14 parts per thousand (°/oo) and died after exposure to 18°/oo. Although this fish could not have lived in normal sea water (35–38°/oo), it was evidently capable of tolerating a moderate degree of brackishness. In areas of high rainfall, the diluting effect of the discharge from even moderate-sized rivers (especially in bays or enclosed arms of the sea) can extend fair distances offshore. Given a zone of tectonic instability and changing shorelines, it appears possible that a fish such as this one, tolerant of moderate salinity, could in the course of time progress slowly—always in an estuarine environment—along the Lesser Sunda chain. This route would provide a connection between the eastern shore of the Banda Sea (with access thence to or from Ceram and the other Moluccas, the Papuasian mainland and Australia) and the Sunda shelf, while by-passing Celebes.

Other freshwater fishes customarily assigned to the primary division on taxonomic grounds have also been found to show limited salt toleration. An example is provided by the spiny eel *Macrognathus aculeatus* which has been taken in brackish water within its principal continental Asian range (Sufi 1956). This is one of three Oriental Mastacembelidae represented in museum collections by specimens purportedly from Ceram and/or the other Moluccas generally (Sufi 1956; cf. Fig. 7.2, above). These records have not been admitted by de Beaufort (in Weber and de Beaufort (1911) vol. 11) and may be invalid. However, should further collecting substantiate this distribution, immigration via the Lesser Sunda islands would once again be implicated.

Environment-induced faunal changes

The distribution of stegodonts demonstrates the availability of a route open to large herbivorous mammals that connected Java, the Lesser Sundas, and Celebes at some time past, probably early in the Quaternary. The close similarity between examples from Celebes, Timor, and Flores led Hooijer (1975) to consider that a single interbreeding population of *S. sompoensis* occupied this area, for which he adopted the name 'Stegoland'. The relations of the Mindanao pigmy stegodont to this population are not clear. There is geographical evidence for a direct land connection between Celebes and the Lesser Sunda Islands (Chapter 4). Further fossil evidence is needed to see whether migration was from the west through Borneo or the east through the Banda Arcs (cf. the Mount Lompobattang flora, Chapter 4). Unfortunately the matter remains enigmatical and no relevant fossiliferous strata are known.

The presence of fossils of stegodonts and extinct types of comparatively large murid rodent in Flores and Timor emphasizes the conclusion, drawn in Chapter 5 from other evidence, that past climates—and hence former vegetation cover and other significant features of the environment—were very different from those prevailing today. The time span covered is great. Stegodont remains have been identified in deposits in Java of lower, middle, and upper Pleistocene age (see Medway (1972) for a summary). Postulated low sea levels during glacial periods presumably facilitated widespread dispersal within the Sunda region. The subsequent extinction of stegodonts, along with other members of the Pleistocene megafauna, is attributable to environmental changes. Likely factors include the dimunition of the area of habitable land at times of high sea level and alteration of the habitat as a consequence of climatic change, perhaps in conjunction with effects of the animals' own activities. The present-day climatic progression within Java, from comparatively non-seasonal and humid in the west to more markedly seasonal and relatively more arid in the east, has been adduced as an explanation of the impoverishment of the vertebrate faunas of the eastern part of the island compared with the western. Diminished island area and still more pronounced seasonality undoubtedly explain much of the further impoverishment of the faunas of the Lesser Sunda Islands where, as in eastern Java, in post-Pleistocene times the natural vegetation has proved very susceptible to degradation, particularly by human interference. In many places the original forest has now been replaced by alang-alang (*Imperata*) grassland. The observations of Lincoln (1975) stress that existing climatic factors are reflected not only in the composition of faunas but also in the relative abundance of species held in common by different islands in the Lesser Sunda chain.

Archaeological evidence has demonstrated the extinction of species of 'giant' rats since the early Holocene in Flores and Timor. Many species of other terrestrial vertebrate groups have probably also disappeared from these islands during this period without leaving adequate fossil records. In modern times, the pervasive Indo-Australian 'tramps' have invaded. Present faunas undoubtedly represent contemporary equilibria reflecting the current operation of factors of general biogeographical significance, including island area, habitat diversity, distances from the principal sources of immigrants, etc. The Lombok Strait does not mark any decisive barrier or faunal boundary. A similar interpretation is equally applicable to the Philippines. We thus see that the northern and southern extremities of Wallace's line were based on assessments of faunal differences which are not borne out by closer analysis.

The role of 'Wallacea' as a centre of radiation

Among birds at least two families, the bee-eaters, Meropidae (24 species, 7–8 genera, Old World tropics and warm temperate zone) and the kingfishers, Alcedinidae (87 species in some 14 genera) attain their greatest diversity and richness in Malesia and have apparently spread east and west from this region which is identified as the likely site of their origin (Fry 1969, 1980*a*). Similar evidence suggests that, among mammals, the true rats, *Rattus*, also originated in Malesia (Misonne 1969). *Rattus* is well represented by separate distinctive assemblages of species in Sundaland, Celebes, and Papuasia. The species present in Celebes were held by Misonne to be 'rather primitive', a consideration which perhaps encouraged Groves (1976) to particularize Celebes as the possible site of the origin of the genus.

Fry (1980*b*) has envisaged that, as the gap between the Laurasian and the Gondwanan plates closed during the Miocene, 'the complicated topography of the region, and the changes in area and outline of subaerial land brought about by subsequent variations in sea-level, produced what must have been ideal circumstances for the multiplication of species'. At the present day, with the sea within a few metres of its highest Pleistocene level, much of this land is now submerged, yet at the time of its formation it had almost continental dimensions. These bird families may date their origin to the Miocene, though fossil evidence on this point is lacking, and the genus *Rattus* plausibly has existed since the late Miocene or early Pliocene (Misonne 1969). The region now termed 'Wallacea' originated from about the mid-Miocene onwards, so evolution of these groups under the conditions visualized by Fry is geographically possible.

However, many of the species of vertebrate found in Malesia today are likely to have evolved more recently, i.e. during the Pleistocene (e.g. the fishes *Rasbora*, Fig. 7.4). The vertebrate faunas of the smaller islands have certainly changed composition, perhaps radically, since the close of the Pleistocene. 'Wallacea' has had a turbulent geological history (Chapter 4) and the present disposition of exposed land is not as old as the region itself. It is concluded that, although local species radiation has undoubtedly occurred in Celebes (and elsewhere in Malesia), there are probably insufficient grounds to conclude that any portion of this island or 'Wallacea' as it exists today can be identified as the site of ultimate origin of any major taxon.

8 WALLACE'S LINE AND SOME OTHER PLANTS

T. C. Whitmore

Examples are discussed of genera and families of plants which have distinctly Sundaic or Papuasian concentrations within present day Malesia. In addition some groups appear to have reached the region from both directions. The problem remains as to how groups have come to have bihemispheric, north and south, ranges meeting or approaching only in Malesia, as it seems improbable that a present day global range from an origin in Malesia could have evolved in the short period of 15 Ma since Malesia was created. This is the challenge Wallace's line still presents to biogeographers.

In Chapter 6 we showed that the geological history of the Malay archipelago is reflected in the geographical ranges of different genera within the single family Palmae which, like many higher plants, have evidence of slow evolution at the generic level. There is evidence that different groups of palms have arrived in Malesia from the Laurasian north and Gondwanic south and a few groups may have arrived from both directions. There has been intermingling to different degrees. Wallace's line has reality for palms because they disperse slowly and the palm flora has not become homogenous. Wegener himself recognized that Wallace's line signified the convergence of Australia on south-east Asia in his discussion of zoological evidence for continental drift (Wegener 1924 and discussion in Hallam 1967).

There have been several attempts to relate the ranges of plants within the Malay archipelago to the geological history of the region. The earlier endeavours were hampered by disagreement and later ones by lack of precision in the palaeogeographical interpretation. Van Steenis (1950) gives a review of the earliest ideas and has himself made major contributions especially regarding mountain plants (see Chapter 5) and those of dry climates (van Meeuwen *et al.* 1961; van Steenis 1979). It was H. J. Lam in the mid-thirties who first espoused Wegenerian views and propounded collision of Australia/New Guinea with western Malesia in the late Tertiary (Lam 1934). This proposition was so far ahead of accepted botanical dogma, reaching back to such nineteenth-century botanical giants as Bentham, Diels, and Hooker, that he soon abandoned the idea (1938), later to be supported in his recantation by van Steenis (1950), Takhtajan (1969), and Smith (1970). But as modern plate tectonic theory has developed in the last decade, the idea of a Laurasian/Gondwanic collision has reappeared in papers by Schuster (1972, 1976), Raven and Axelrod (1972, 1974), Walker (1972, 1981), Aubréville (1975), Johnson and Briggs (1975), van Balgooy (1976), Raven (1979), and most recently by van Steenis himself (van Steenis 1979).

In this final chapter we examine a small selection of plant groups whose present day ranges within Malesia we believe, like the palms, clearly reflect the geological history. Previous chapters have shown that such interpretation is fraught with difficulties. The field of biogeography is strewn with the bodies of similar attempts, mostly killed by later discoveries on distribution, on relationships, or, most important, on palaeogeography. It is precisely because the palaeogeography has now been more or less settled that this book has been written. The examples are chosen, with the other difficulties firmly in mind, from groups which are well known and, where appropriate, the interpretations are illuminated by consideration of evolutionary relationships within them as was the case with the palms. Specialists will be able to multiply the

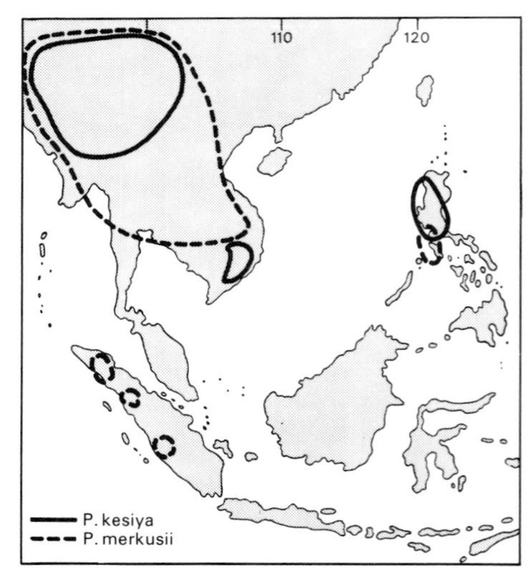

Fig. 8.1. *Pinus* in Malesia. The ranges of *P. kesiya* and *P. merkusii* (Critchfield and Little 1966; Cooling 1968).

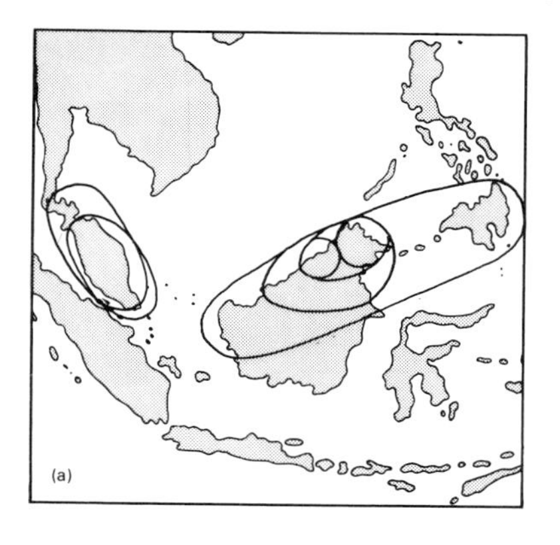

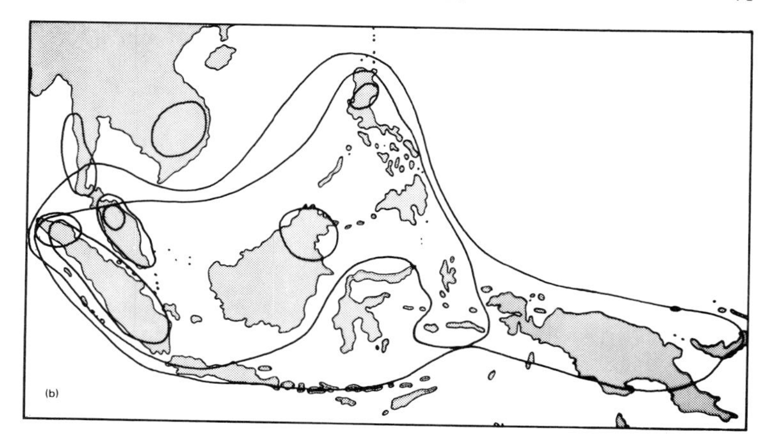

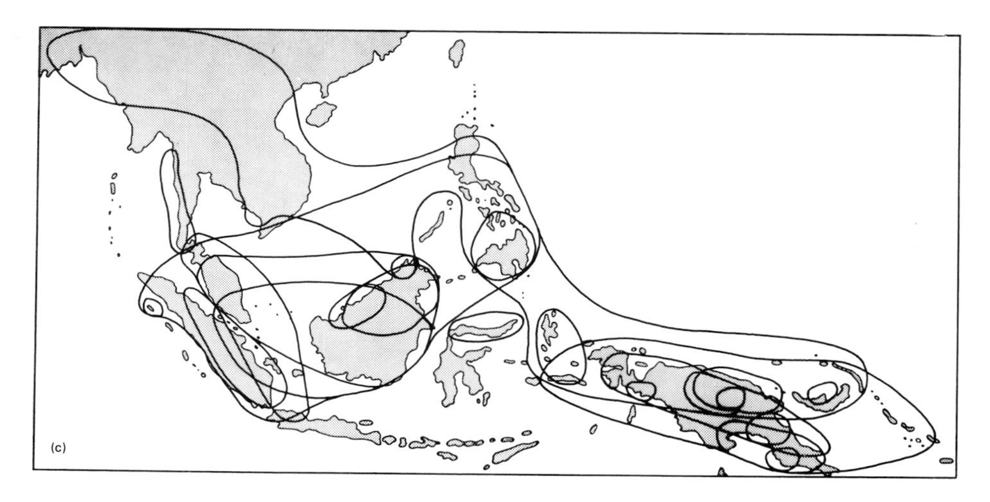

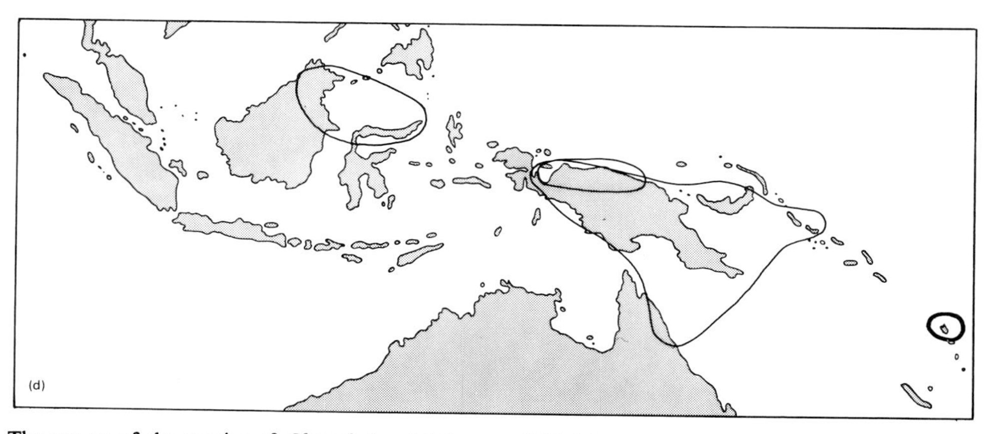

Fig. 8.2. The ranges of the species of *Chisocheton*, Meliaceae, divided into their natural subgroups. (a) Section *Clemensia*. (b) Section *Dasycoleum*. (c) Section *Chisocheton*. (d) Section *Rhetinosperma*. (Data of Mabberley (1979).)

examples many times over. The examples chosen are families or genera most of whose species are rain forest trees. Plants of seasonal climates and high mountains exhibit patterns which strongly reflect climatic history. They are less suitable for the present demonstration and have been briefly discussed in Chapter 5.

LAURASIAN GROUPS

The maps, Figs 8.1 to 8.4, illustrate the ranges of a number of families and genera concentrated in western Malesia and which the evidence suggests have a Laurasian origin.

Conifers

It is well established that the Coniferae divide into two great classes, of northern and southern hemisphere range respectively (Florin 1962). Both classes have representatives in Malesia and examples of each are considered here. Amongst the northern conifers only *Pinus* (Fig. 8.1) extends into Malesia at the present day. In addition, pollen of *Picea* and *Pinus* has been found in north-west Borneo from the Eocene to the Pliocene (Muller 1966).[1]

Pinus is represented by two species with partly overlapping ranges, *P. merkusii* in Sumatra and the Philippines and *P. kesiya* in the Philippines only. For both species the Malesian stations are the extremity of a wide range in seasonally dry continental south-east Asia. Indeed, within Malesia they are both confined to slightly seasonally dry climates and are pioneer species colonizing open ground after fire or landslip and are most persistent on shallow rocky soils. There is now considerable knowledge of their ecology, see van Steenis (1957), Kowal (1966), Cooling (1968), Whitmore (1975), and Stein (1978). Like many of the other subtropical species of *Pinus* these two species are of considerable forestry importance, though less so than their middle American counterparts *Pinus caribaea* and *P. oocarpa*.

Chisocheton

Chisocheton is a genus of fifty-one species of small to medium-sized trees of tropical rain forest. It is a member of the mahogany family Meliaceae though not itself producing useful timber. Like most of the family, *Chisocheton* species have pinnate leaves and the genus is well known to tropical botanists because in most species the leaf continues to grow long after the bud has unfurled. with circinnate vernation of the rachis tip. A recent study of the genus and its evolution (Mabberley 1979) shows that there are four sections. The most primitive section, *Clemensia* is confined to west Malesia (Fig. 8.2a). Section *Dasycoleum* is concentrated in west Malesia but with two species ranging to the east (Fig. 8.2b). Section *Chisocheton* has two series; one is centred on west Malesia, the other, consisting of very restricted endemic species plus one widely ranging polymorphic species, is eastern, centred on New Guinea (Fig. 8.2c), but with four linking species in the first series in the west plus an incompletely known species (*C. aenigmaticus*) from Simalur island tentatively placed here. The fourth section, *Rhetinosperma*, which is probably derivative from section *Dasycoleum* is concentrated in Papuasia (Fig. 8.2d). The ranges of the four sections of *Chisocheton* (Fig. 8.2), together with the insight into their evolutionary relationship, suggest that this is a genus of Sundaic origin which has diversified and speciated in the course of migrating into Papuasia, northern Australia and the Melanesian islands.

[1] The extraordinary flora recorded by Zaklinskaya (1978) from a marine core obtained off south-west Timor by Glomar Challenger in 1972 included taiga elements as well as Laurasian conifers. These must be contaminants (J. Muller, personal communication).

Dipterocarpaceae

Dipterocarpaceae epitomize the lowland tropical rain forests of Sundaland. The family in fact ranges from Africa (the endemic subfamily Monotoideae) through India, south China, Indo-China and Malesia, to the D'Entrecasteaux Islands north-east of Papua; but it is only in Malesia, west of Wallace's line, that it attains family dominance, the most abundant tree family of the upper part of the forest canopy, and at present being vigorously 'mined' for timber.

Sumatra, Malaya, and Borneo have ten genera (three endemic) and over 280 species. The Philippines, where dipterocarps still dominate, has only six genera and about 39 species (Rojo 1979). In Java, the Lesser Sunda Islands, Celebes, the Moluccas, and New Guinea there are far fewer dipterocarps. Celebes has 2 genera and 45 species, the Moluccas have 4 genera each with a few species and New Guinea has only 3 genera, *Anisoptera* and *Vatica* each with one species and *Hopea* with 11 closely-knit endemic species. The pattern within Malesia (Fig. 8.3) is of strong concentration in the Sundaland rain forests with limited extension east of Wallace's line as far as Papuasia and northwards in the seasonally dry continent. The Asian subfamily, Dipterocarpoideae, is also represented in Seychelles (one species) and Ceylon (7 genera, 45 species) which each have small endemic genera. This pattern is similar to that already noted for several groups of palms and likewise suggests migration from an early centre, on West Gondwanaland.

Magnoliaceae

Six genera of Magnoliaceae occur in Malesia and four

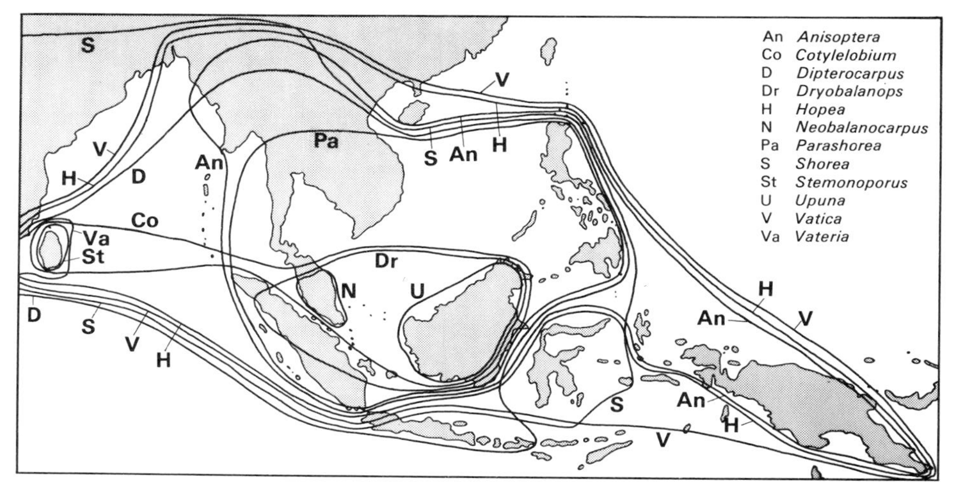

Fig. 8.3. The ranges of genera of Dipterocarpaceae in Malesia and adjacent Asia (with the assistance of P. S. Ashton, personal communication).

more are endemic to the region of south China and northern Indo-China (Fig. 8.4). Within Malesia *Elmerrillia* is endemic from east Borneo to New Britain. There is a very marked concentration in western Malesia. Malaya and Sumatra both have five genera. Borneo has three genera, Philippines two, Celebes one, and New Guinea two. Magnoliaceae as a family are strongly concentrated in sub-tropical east Asia. They extend into Malesia from this northern bastion becoming progressively less well represented eastwards and supplemented there by two endemic genera. Magnoliaceae are accepted by most botanists to be a primitive family. The distribution contrasts strongly with that of Winteraceae, also regarded as primitive, which is described below. The two are further discussed at the end of the chapter.

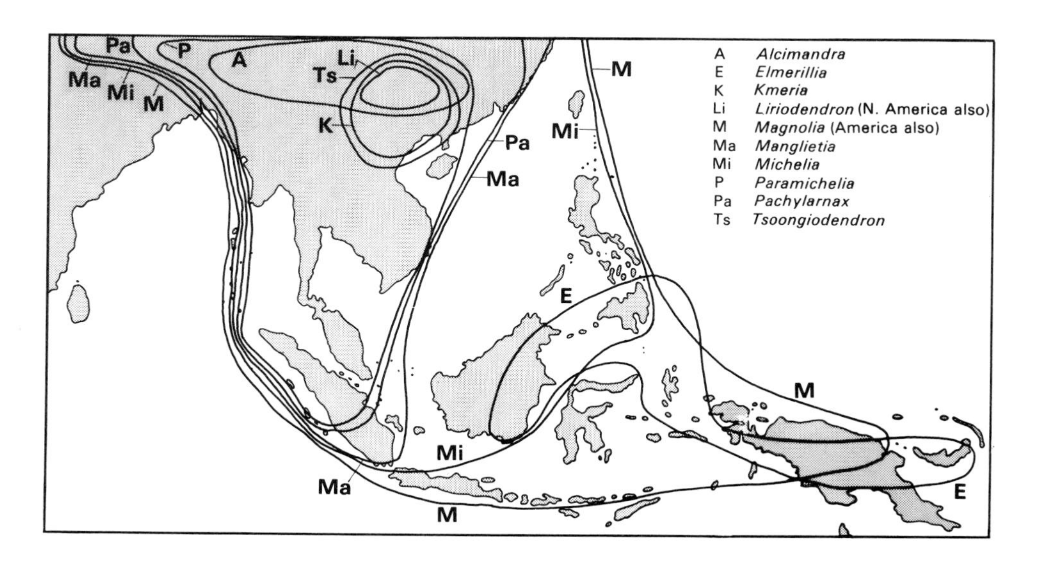

Fig. 8.4. The ranges of the genera of Magnoliaceae in Malesia and south-east Asia (after Schuster 1972, Fig. 31 and in accordance with Keng 1978*b*).

PAPUASIAN GROUPS

The maps Figs 8.6 to 8.9 show a number of families concentrated in Papuasia or the islands of the south-west Pacific, with weaker representation in Sundaland. These groups appear to enter the Malay archipelago from Gondwanaland.

Fig. 8.5. *Phyllocladus hypophyllus*, Podocarpaceae, the 'Celery Pine' (Dakkus 1925).

Phyllocladus

The 'Celery Pine' *Phyllocladus* is a small genus of the big southern conifer family Podocarpaceae. There are three species in New Zealand, one in Tasmania and one, *P. hypophyllus*, in montane rain forests in Malesia extending west as far as Philippines and eastern Borneo (Figs. 8.5 and 8.6). The range is thus strongly Gondwanic. It is particularly interesting that in pollen profiles from north-west Borneo, which extend from the Oligocene (30 Ma) onwards, *Phyllocladus* pollen appears rather suddenly at the Plio-Pleistocene boundary, 2–3 Ma (Muller 1966). *Podocarpus imbricatus* pollen also appears at about the same time. It will be recalled that the collision between Laurasia and Gondwanaland is dated at *c.* 15 Ma.

Araucariaceae

Agathis, the Kauri Pines, and *Araucaria*, the Bunya, Hoop and Paraña Pines, and the Monkey Puzzle, form a distinctive, taxonomically isolated, family amongst the conifers which have a southern hemisphere range. *Agathis* extends north of the equator in Malesia. *Araucaria* is entirely confined to the southern hemisphere, attaining about 0°30′S in Waigeo. Both genera are highly prized for their valuable timber (Whitmore 1977*b*, 1980; Ntima 1968).

Agathis occurs throughout the Malay archipelago and in the south-west Pacific (Fig. 8.7) where it is represented by several mainly allopatric species (Whitmore 1980). Only in the rain forests of north Queensland and New

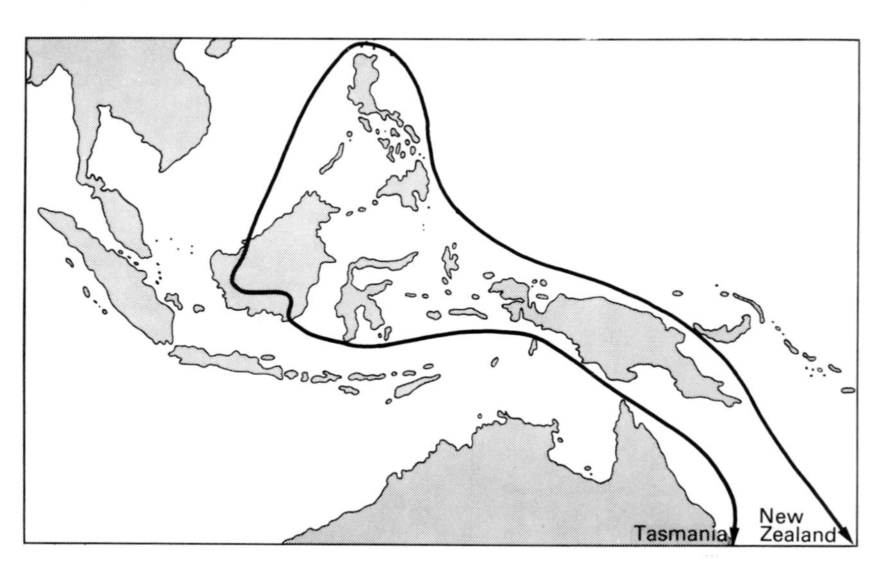

Fig. 8.6. The range of *Phyllocladus*, Podocarpaceae (Keng 1978*a*).

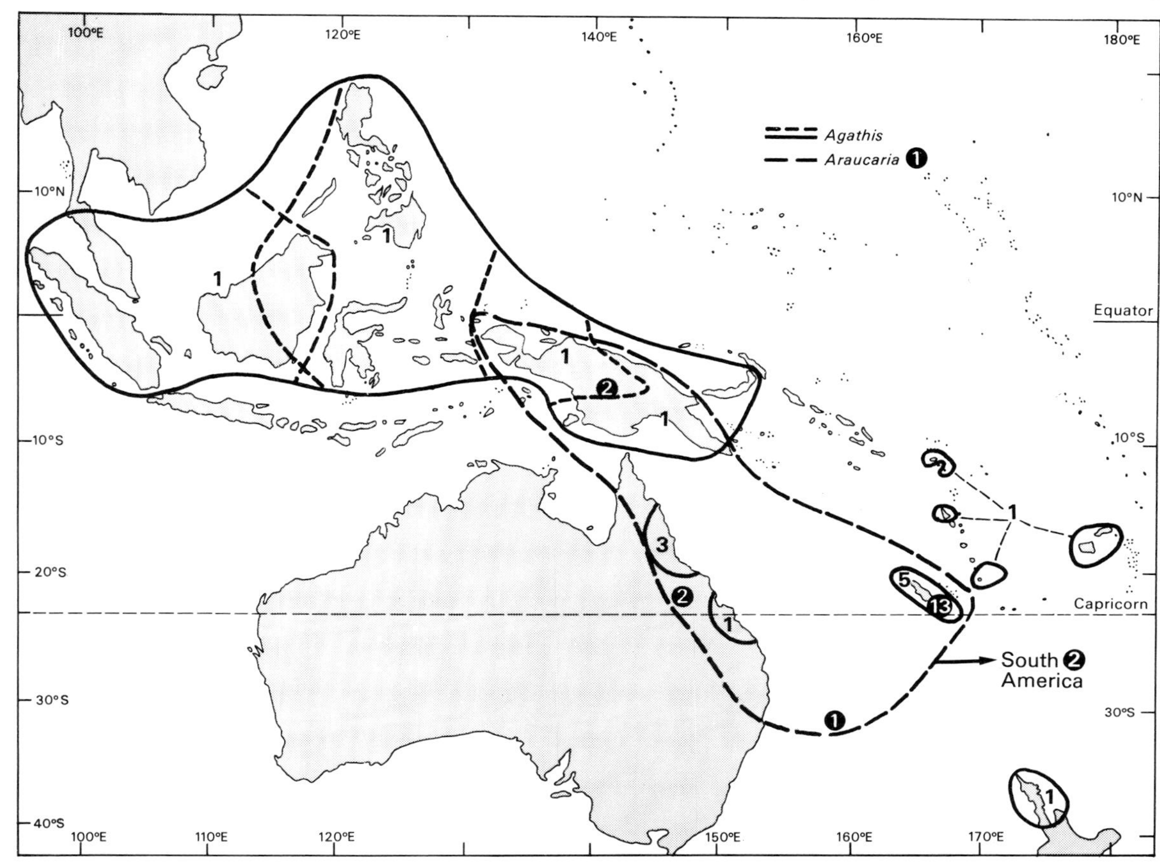

Fig. 8.7. The Old World ranges of *Agathis* and *Araucaria*, Araucariaceae. (Partly after Whitmore 1980.)

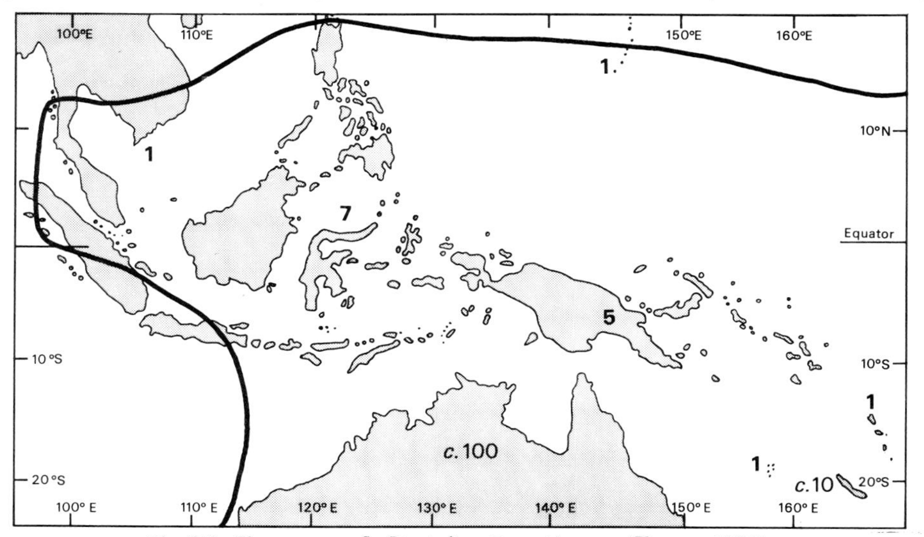

Fig. 8.8. The range of *Styphelia*, Epacridaceae (Sleumer 1964).

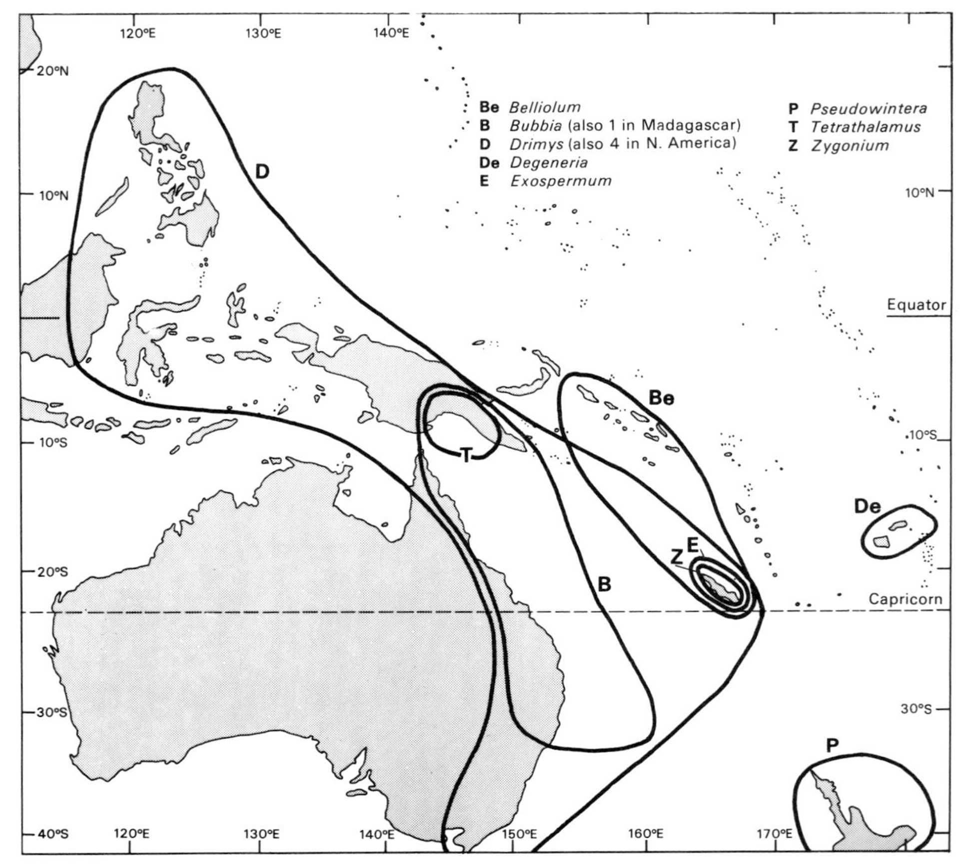

Fig. 8.9. The ranges of the genera of Winteraceae in Malesia and the south-west Pacific. *Degeneria* is often segregated as a separate monotypic family.

Caledonia are there sympatric species, three and five respectively. Careful analysis of details of the habitat, population structure, and taxonomic relationships leads to the conclusion that *Agathis* has extended from these two centres (Whitmore and Page 1980).

Araucaria has not been investigated in such detail. It occurs in South America (two allopatric endemic species), and Norfolk Island (one endemic species), Australia (two species, one endemic), New Guinea and Waigeo (two species, one endemic), and New Caledonia and the Loyalty Islands (thirteen endemic species) (Fig. 8.7). The centre of Araucariaceae is clearly on parts of the Gondwanic fragments south of Malesia. The range within Malesia is Gondwanic.

Styphelia

This is a genus of sclerophyllous shrubs and small trees of the Epacridaceae. Like the family it is strongly centred in Australia. A few species of *Styphelia* occur in the Malay archipelago (Fig. 8.8) where they are confined to acid, sandy or peaty, soils, for example in lowland heath forest and upper montane and subalpine rain forest, sites subject to either periodic water stress or extreme deficiency of mineral nutrients or both (Sleumer 1964; Whitmore 1975; van Steenis 1979). Within Malesia *Styphelia* represents a pattern of inter-digitation of a characteristically Australian floristic element into the Malesian flora on poor sites which is also shown by other Australian-centred genera, for example *Baeckia*, *Leptospermum*, and *Melaleuca* of the Myrtaceae. This was first pointed out by Richards (1943).

Epacridaceae are closely related to Ericaceae which they replace in Australia.

Winteraceae

This family of primitive flowering plants is strongly

centred on the islands of the south-west Pacific (Fig. 8.9) (Smith 1943; Vink 1970). Two genera have outposts elsewhere, one species of *Bubbia* occurs in Madagascar and four of *Drimys* in South America.

Within the Malay archipelago the centre is eastern New Guinea, with three genera, of which *Drimys* (Fig. 8.10) extends westwards to cross Wallace's line into the Philippines and eastern Borneo.

Winteraceae are the southern counterpart to Magnoliaceae.

Fig. 8.10. *Drimys piperita*, Winteraceae, at the western limit of the family range in Borneo at Gunung Mulu National Park.

MALESIAN GROUPS OF APPARENTLY DUAL ORIGIN

There is not much doubt that the examples of families and genera given above reached Malesia from either Laurasia or Gondwanaland. The evidence is their present day distribution patterns backed up by what is known about relationships. Any alternative hypothesis for their Malesian ranges would be much more extravagent and an offence to Ockham's razor. It is much more difficult to identify families which have reached Malesia from both directions as this presupposes a very thorough understanding of the group in question.

Proteaceae

Proteaceae have been subject of a recent, highly detailed investigation (Johnson and Briggs 1975) and also enumerated for Malesia (Sleumer 1955). The family is confined to the southern hemisphere and strongly concentrated on Australia and the southern tip of Africa. The ranges of seven Australian genera extend into the Malay archipelago (Fig. 8.11) Here they are confined to New Guinea, except *Grevillea* and *Macadamia* which reach seasonally dry parts of Celebes and *Helicia*, rain forest trees, which ranges as far west as Ceylon and continental south-east Asia. One genus, *Finschia*, is endemic to east Malesia and the south-west Pacific. This pattern is similar to that of other families which occur in Malesia from a Gondwanic origin. However, the genus *Heliciopsis* presents an anomaly. This is endemic to Malesia and the adjacent continent west of Wallace's line (Fig. 8.11). Johnson and Briggs believe that *Heliciopsis* could have reached south-east Asia from West Gondwanaland, by the 'Noah's Ark' medium of India (in which it subsequently died out). Proteaceous pollen is known from the Eocene in India. Fuller study of relatives in Madagascar, Queensland, and New Caledonia is now needed.

Nastus

Nastus, a small genus of about fifteen species of bamboos which scramble through the canopy of tropical forests, occurs in Java, Sumba and Flores, islands on the fringe of Laurasia, and possibly Sumatra too. It is also in Gondwanic New Guinea and the Bismarck and Solomon archipelagos. In addition *Nastus* occurs on the south Indian ocean islands of Madagascar and Réunion (Fig. 8.12). This range is similar to that of several palms described in Chapter 6 and likewise suggests a West Gondwanic origin followed by arrival in Malesia via both Laurasia and Australia/New Guinea (Soenarko-Dransfield, unpublished).

Fagaceae

The Fagaceae are divided into three subfamilies. The oaks and sweet chestnuts, subfamilies Quercoideae and Castanoideae, are richly developed in west Malesia, with four genera (Fig. 8.13) (Soepadmo 1972). *Trigonobalanus* has two Asian species, *T. doichangensis* in the mountains of north Thailand, and *T. verticillata* in small areas of Malaya, Sumatra (Anon. 1977), Borneo and Celebes. *Quercus* has 19 Malesian species and is in Sumatra, Malaya, extreme west Java, Borneo and Palawan. *Lithocarpus* extends east to New Guinea. There are 104 species in Malesia, most of them in Sundaland, but with 32 in Philippines (19 endemic), four in Celebes, one in Moluccas, and 16 in New Guinea. *Castanopsis* has 34 Malesian species with the same overall range and pattern of eastwards diminishment (5 Philippines, 2 Celebes, 1 Moluccas, 1 New Guinea). Outside south-east Asia *Quercus* occurs throughout Laurasia and in North, Central, and north-west South America. *Castanopsis* and *Lithocarpus* are both found in east Asia and each also has one species in south-west

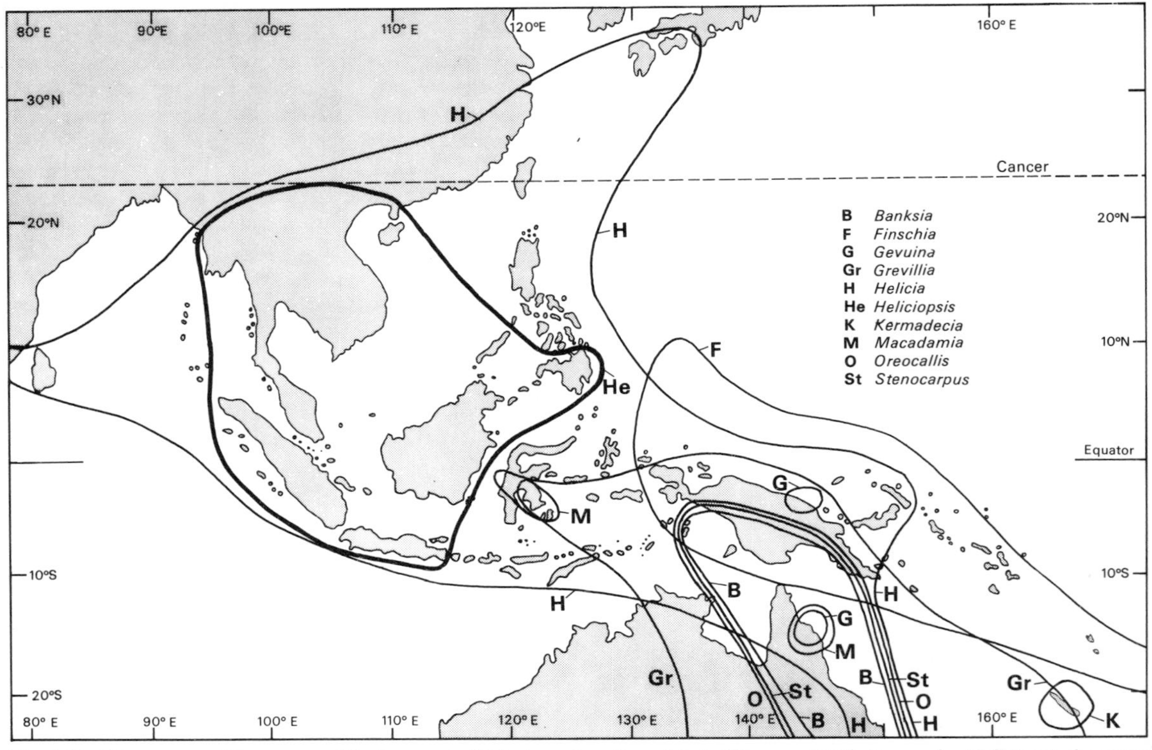

Fig. 8.11. The ranges of the genera of Proteaceae in Malesia and south-east Asia (Sleumer 1955). Except for *Heliciopsis* the centre is eastern.

Fig. 8.12. The bamboo *Nastus borbonicus* on Réunion in the south Indian Ocean. Disturbed *Acacia heterophylla* forest *c.* 1200 m elevation, between Maido and Guillaume. The genus also occurs in west Malesia and the Bismarck and Solomon archipelagos of the south-west Pacific.

North America. *Trigonobalanus*, which was only recognized as a genus in 1962 and believed endemic to south-east Asia, was in 1979 discovered in the northern Andes, in Colombia (Logano, Hernández, and Henao 1979). Thus the pattern of these subfamilies is strongly Laurasian, similar to the examples already discussed.

The third subfamily, Fagoideae, occurs in east Malesia where it is represented by the southern beech, *Nothofagus* (Fig. 8.13). *Nothofagus* has a very complete and long fossil record and highly distinctive pollen. It has been the subject of considerable phytogeographic discussion (van Steenis 1971, 1972*b*; Raven and Axelrod 1974; Schuster 1972). At the present day it is found in montane rain forests in New Guinea and New Britain where sixteen species occur, often dominant and sometimes as consociations. *Nothofagus* is also found in lowland, temperate evergreen forests of southern Australia, New Caledonia, Tasmania, and New Zealand as well as in temperate forests of South America. Its absence from northern, tropical Australia must result from extinction during dry periods of the Pleistocene as discussed in Walker (1972). The range of *Nothofagus* suggests a genus which has reached Malesia on the Australia/New

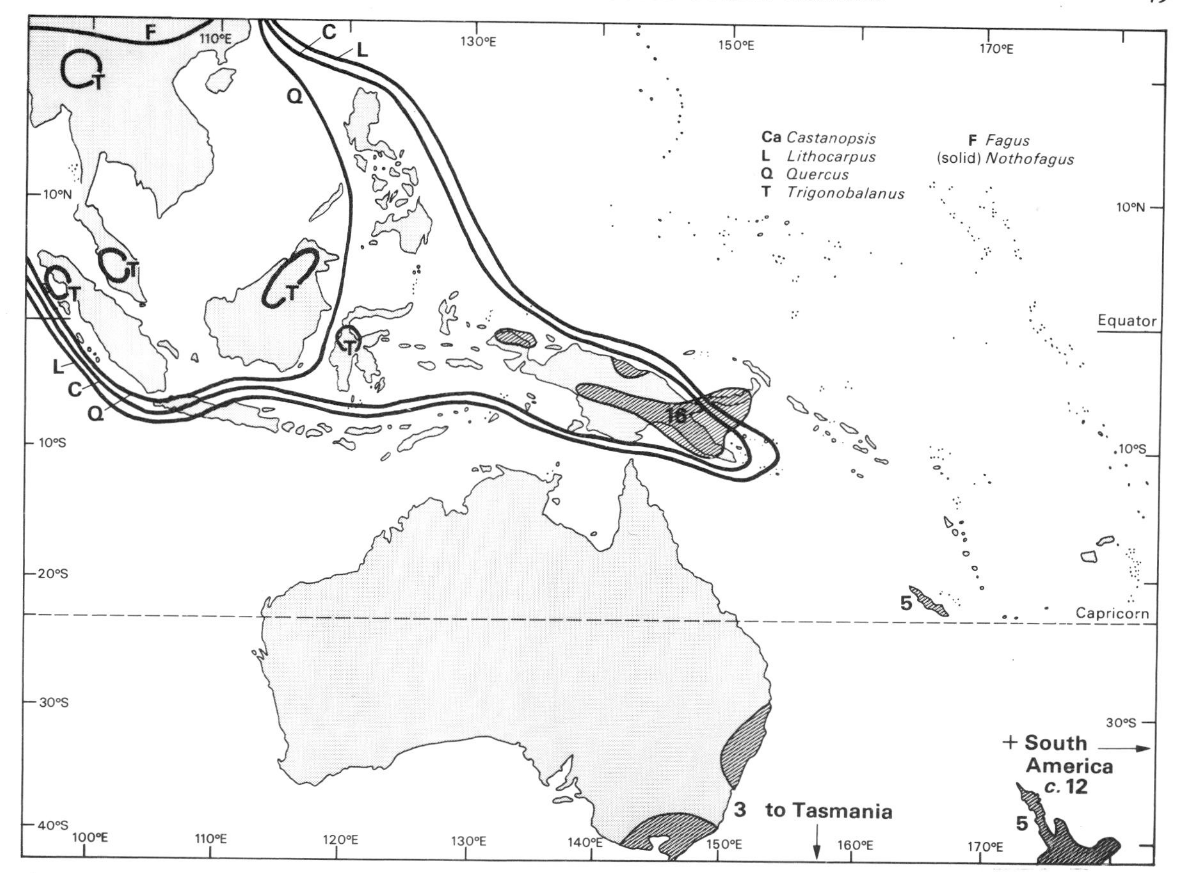

Fig. 8.13. The ranges of the genera of Fagaceae in south-east Asia, Malesia, and the south-west Pacific (Soepadmo 1972): all except *Nothofagus* are western centred.

Guinea fragment of Gondwanaland. No species occur in the tropical lowlands and it is suggested that it survived on the northern fringe of Gondwanaland, as that drifted northwards, by migrating to the mountains where it was also protected from the effects of dry periods during the Pleistocene. The superficial similarities in climate and forests between wet tropical mountains and wet temperate lowlands have often been remarked (Whitmore 1975; Grubb 1977).

The ranges of the genera of Fagaceae in the Malay archipelago thus suggest a dual origin there but the converse of the Proteaceae: most genera Laurasian plus one Gondwanic, entry mainly from the north but also from the south.

THE REMAINING ENIGMA

There is, however, in Fagaceae a complicating factor which instances one of the remaining great puzzles of phytogeography. This we now mention although in its full extent it is beyond the scope of the present book. *Nothofagus* is the southern genus of subfamily Fagoideae. The true beech *Fagus* is the other genus and its range is entirely in the northern hemisphere, extending to southern China (Figs 1.1, 8.13).

South-east Asia and Malesia is the region where the ranges of the two genera of this undoubtedly natural subfamily come closest to each other.

Magnoliaceae of the northern hemisphere (Fig. 8.4) and Winteraceae of the southern hemisphere (Fig. 8.9) are both families widely believed to be primitive and related and their ranges overlap in Malesia.

Ericaceae, mainly of the northern hemisphere, is richly represented throughout Malesia where it overlaps with *Styphelia* of the closely related southern hemisphere family Epacridaceae (Fig. 8.8).

Two other pairs of families which overlap in Malesia

are Staphyleaceae and Cunoniaceae, and Saxifragaceae and its very close relative Escalloniaceae. These pairs are northern and southern respectively.

Botanists have attempted to account for these bihemispheric pairs of related groups. Now that there is a general understanding of global continental movements and with the discovery that many families appear to have migrated out from West Gondwanaland, South America plus Africa (Raven and Axelrod 1974) the discussions have at least a few fixed points. But in the absence of either an adequate fossil record or an accepted view of evolutionary relationships within the group under discussion, or of both, controversy is likely to continue to rage (Takhtajan 1969; Smith 1970; Schuster 1972, 1976; Raven 1979; van Steenis 1979). Malesia is the region where these northern and southern groups today come closest to each other in range and in some cases they actually overlap. The archipelago has only been in existence for 15 Ma, since the mid-Miocene; and the explanation we have offered here is that within Malesia there has been convergent migration and some further evolution of species since then of genera which were already in existence at the time of collision.

There is scanty knowledge on the evolutionary rates of genera of flowering plants but what is known of generic interrelationships and ranges and the evidence of micro- and macro-fossils (Muller 1974) makes it unlikely that pairs of families could have differentiated within Malesia since the collision and then migrated to attain their present world range. It is similarly unlikely that they could have evolved on the margin of either Laurasia or Gondwanaland prior to the collision and have spread rapidly through the other hemisphere in the last 15 Ma.

In order to explain these enigmas van Steenis (1979) has postulated that there was in the mid-Cretaceous or earlier a land mass connecting Laurasia and Gondwanaland in the region of present-day Malesia, a cradle in which primary differentiation of families (e.g. Magnoliaceae/Winteraceae) or genera (e.g. *Fagus*/*Nothofagus*) took place. This postulate assumes that present day distribution patterns do indeed reflect primary differentiation not substantial extinction and evolution in secondary centres. We showed in Chapter 3 that there is geological evidence for a contact between Gondwanaland and Laurasia about the late Jurassic or early Cretaceous when a shard of north-west Australia separated and drifted north. This shard has not been found though it may lie in Assam. At present it has only a phantom existence and may have been entirely submarine. Its relationship, if any, with the Ninetyeast ridge, another poorly known, and sometimes subaerial, north-south link, is unknown. Fossil pollen from the Ninetyeast ridge is exclusively Australasian and partly at least wind-borne (Kemp and Harris 1975).

Could these lands have been cradles for major differentiation in flowering plants, rather than just bridges or rafts for contact between Gondwanaland and Laurasia, similar to but much earlier than the 'Noah's Arks' of India and the Malay archipelago? To answer this we must await further details on the date these lands existed and moved and on the early evolution of flowering plants.

Our present knowledge of the biota of the region of Wallace's line throws this much larger biogeographical problem into sharp focus and raises new questions for geological study. Painstaking and critical biological research is also needed on the Tertiary evolution and migration of other related groups of plants and animals. There is too little fossil evidence.

The revolution in the earth sciences has led to a total reappraisal of the biogeography of the Malay archipelago but Wallace's line remains today, as for the past 120 years, a cogent influence, powerfully able to generate hypotheses subject to further test. It is still a challenge to biogeographers and geologists.

BIBLIOGRAPHY

Abbott, M. J. and Chamalaun, F. H. (1978). New K/Ar age date for Banda Arc volcanics. *Publications of the Institute of Australian Geodynamics, South Australia* 78/5, 1–34.

Anderson, J. A. R. and Muller, J. (1975). Palynological study of a Holocene peat and a Miocene coal deposit from N.W. Borneo. *Review of Palaeobotany and Palynology* **19**, 291–351.

Anon. (1977). *Flora Malesiana Bulletin* **30**, 2767.

Aubréville, A. (1975). La Flore Australo-Papou. Origine et distribution. *Adansonia* **15**, 159–70.

Audley-Charles, M. G. (1978). The Indonesian and Philippine archipelagos. In *The phanerozoic geology of the world, II the Mesozoic, A.* pp. 165–207. Elsevier, Amsterdam.

—— (1980). Geometrical problems and implications of large scale overthrusting in the Banda Arc—Australian margin collision zone. In *Thrust and nappe tectonics* (ed. K. McClay and N. J. Price). Geological Society of London Special Publication **9**, 407–16.

—— Carter, D. J., and Milsom, J. S. (1972). Tectonic development of eastern Indonesia in relation to Gondwanaland dispersal. *Nature, London (Physical Science)* **239**, 35–9.

—— and Hooijer, D. A. (1973). Relation of Pleistocene migrations of pygmy stegodonts to island arc tectonics in eastern Indonesia. *Nature, London* **241**, 197–8.

—— and Milsom, J. S. (1974). Comments on a paper by T. J. Fitch 'Plate convergence, transcurrent faults, and internal deformation adjacent southeast Asia and the western pacific'. *Journal of Geophysical Research* **79**, 4980–1.

—— Carter, D. G., Barber, A. J., Norvick, M. S., and Tjokrosapoetro, S. (1979). Reinterpretation of the geology of Seram: implications for the Banda Arcs and northern Australia. *Journal Geological Society of London* **136**, 547–68.

Bain, J. H. C. (1973). A summary of the main structural elements of Papua New Guinea. In *The Western Pacific: island arcs, marginal seas, geochemistry* pp. 149–61. University of Western Australia Press, Perth.

Balgooy, M. M. J. van (1976). Phytogeography. In *New Guinea vegetation* (ed. K. Paijmans) pp. 1–22. Australian National University, Canberra.

Bănărescu, F. (1975). *Principles and problems of zoogeography.* Translated and published for US Department of Commerce by NOLIT, Belgrade.

Barbour, T. (1912). A contribution to the zoogeography of the East Indian islands. *Memoirs of the Museum of Comparative Zoology, Harvard* **44**.

Beaufort, L. F. de (1951). *Zoogeography of land and inland waters.* Sidgwick and Jackson, London.

Beccari, O. (1877). Palme Nuova Guinea. *Malesia* **1**, 9–102.

—— (1908). Asiatic Palms—Lepidocaryeae. The species of *Calamus. Annals of the Royal Botanic Gardens Calcutta* **11**, 1–518.

—— (1918). Asiatic Palms—Lepidocaryeae. The species of the genera *Ceratolobus, Calospatha, Plectocomia, Plectocomiopsis, Myrialepsis, Zalacca, Pigafetta, Korthalsia, Metroxylon, Eugeissona. Annals of the Royal Botanic Garden Calcutta* **12**, 1–231.

Bemmelen, R. W. van (1949). *The geology of Indonesia.* Government Printing Office, The Hague.

Ben Avraham, Z. and Uyeda, S. (1973). The evolution of the China Basin and the Mesozoic palaeogeography of Borneo. *Earth and Planetary Sciences Letters* **18**, 265–76.

Boonsong, L. and McNeely, J. (1977). *Mammals of Thailand.* Kurusapha, Bangkok.

Bowin, C., Purdy, G. M., Shor, G., Johnston, C., Lawver, L., Hartono, H. M. S., and Jezek, P. (1980). Arc-continent collision in the Banda Sea region. *Bulletin of the American Association of Petrologists and Geologists* **64**, 868–915.

Britton, M. R. (1954). A revision of the Indo-Malayan freshwater fish genus *Rasbora. Monographs of the Institute of Science and Technology. Manila* **3**. Reprinted (n.d.), with colour supplement, as: *Rasbora,* Tropical Fish Hobbyists.

Burret, M. (1942). Neue Palmen aus der Gruppe der Lepidocaryoideae. *Notizblatt des Botanischen Gartens und Museums zu Berlin* **15**, 728–55.

Carter, D. J., Audley-Charles, M. G., and Barber, A. J. (1976). Stratigraphical analysis of island arc-continental margin collision in eastern Indonesia. *Journal of the Geological Society of London* **132**, 179–98.

Chamalaun, F. H., Lockwood, K., and White, A. (1976). The Bouguer gravity field and crustal structure of eastern Timor. *Tectonophysics* **30**, 241–59.

Chandler, M. E. J. (1961–4). *The Lower Tertiary floras*

of southern England. British Museum (Natural History), London.

Colbert, E. H. (1973). *Wandering lands and animals*. Dutton, New York.

Cooling, E. N. G. (1968). *Pinus merkusii. Fast growing timber trees of the lowland tropics*, No. 4. Commonwealth Forestry Institute, University of Oxford.

Corbet, A. S. (1943). Considerations based on the Rhopalocerous fauna. *Proceedings of the Linnean Society of London* **154**, 143–8.

Corner, E. J. H. (1966). *The natural history of palms*. Weidenfeld and Nicolson, London.

Critchfield, W. B. and Little, E. L. Jr. (1966). *The geographic distribution of the pines of the world*. United States Department of Agriculture Miscellaneous Publications **991**, 1–97.

Crook, K. A. W. and Belbin, L. (1978). The southwest Pacific area during the last 90 million years. *Journal of the Geological Society of Australia* **25**, 23–40.

Curray, J. R., Moore, D. G., Lawver, L. A., Emmel, F. J., Raitt, R. W., Henry, M., and Kieckhefer, R. (1979). Tectonics of the Andaman Sea and Burma, *American Association of Petrologists and Geologists Memoir* **29**, 189–98,

Dakkus, P. (1925). Naar den boekit Raja in Centraal-Borneo. *Tropische Natuur 1925* 137.

Darlington, P. J., Jr. (1957). *Zoogeography: the geographical distribution of animals*. Wiley, New York.

Diamond, J. M. (1970). Ecological consequences of island colonization by Southwest Pacific birds. 1. Types of niche shifts. *Proceedings of the United States Academy of Sciences* **67**, 529–36.

Dickerson, R. E. (1928). Distribution of life in the Philippines. *Philippine Bureau of Sciences Monograph* **21**.

Dickinson, W. R. (1973). Reconstruction of past arc-trench systems from petrotectonic assemblages in the island arcs of the western Pacific. In *The Western Pacific: island arcs, marginal seas and geochemistry* (ed. P. J. Coleman) pp. 569–601. University of Western Australia Press, Perth.

Dransfield, J. (1976). A note on the habitat of *Pigafetta filaris* in North Celebes. *Principes* **20**, 48.

—— (1977). A dwarf *Livistona* (Palmae) from Borneo. *Kew Bulletin* **31**, 759–62.

—— (1978). Growth forms of rain forest palms. In *Tropical trees as living systems* (ed. P. B. Tomlinson and M. H. Zimmermann). Cambridge University Press.

—— (1979). A manual of the rattans of the Malay Peninsula. *Malayan Forest Records* **29**.

Durden, C. J. (1974). Biomerization: an ecologic theory of provincial differentiation. In *Paleogeographic provinces and provinciality* (ed. C. A. Ross), pp. 18–53. Tulsa, Oklahoma.

Earl(e), G. W. (1845). On the physical structure and arrangement of the islands of the Indian Archipelago. *Journal of the Royal Geographical Society* **15**, 358–65.

Essig, F. B. (1977). The palm flora of New Guinea. A preliminary analysis. *Botany Bulletin* **9**, 1–39. Office of Forests, Division of Botany, Lae, Papua New Guinea.

Fairchild, D. (1943). *Garden islands of the great East*. Charles Scribner's Sons, New York.

Falvey, D. A. (1972). Sea-floor spreading in the Wharton basin (northeast Indian Ocean) and the breakup of eastern Gondwanaland. *Australasian Petroleum Exploration Association Journal* **12**, 86–8.

—— (1974) The development of continental margins in plate tectonic theory. *Australasian Petroleum Exploration Association Journal* **14**, 96–106.

Flenley, J. R. (1979). *The equatorial rain forest—a geological history*. Butterworth, London.

Florin, R. (1962). The distribution of conifer and taxad genera in time and space. *Acta Horti Bergiani* **20**, 121–317, 319–26.

Fooden, J. (1969). Taxonomy and evolution of the monkeys of Celebes. *Bibliotheca Primatologica* **10**, 1–148.

Froidevaux, C. M. (1974). Geology of Misool Island (Irian Jaya). *Proceedings of the Indonesian Petrological Association* (3rd Annual Convention, Jakarta, June 1974), 189–96.

Fry, H. C. (1969). The evolution and systematics of bee-eaters (Meropidae). *Ibis* **111**, 557–92.

—— (1980*a*). The evolutionary biology of the kingfishers (Aedinidae). *Living Bird* **18**, 113–60.

—— (1980*b*). The origin of Afrotropical kingfishers. *Ibis* **122**, 57–72.

Furtado, C. X. (1956). Palmae Malesicae 19. The genus *Calamus* in the Malayan Peninsula. *Gardens' Bulletin Singapore* **15**, 32–265.

George, W. (1964). *Biologist philosopher: a study of the life and writings of Alfred Russel Wallace*. Abelard-Schuman, London.

Gorman, M. (1979). *Island ecology*. Chapman and Hall, London.

Grandison, A. G. C. (1978). Snakes of West Malaysia and Singapore, *Annalen des Naturhistorischen Museums, Wien* **81**, 282–303.

Groves, C. P. (1976). The origin of the mammalian fauna of Sulawesi (Celebes). *Zeitschrift für Säugetierkunde* **41**, 201–16.

—— (1980*a*). Notes on the systematics of *Babyrousa*

(Artiodactyla, Suidae). *Zoologische Mededeelingen, Leiden* **55**, 29–46.

Groves, C. P. (1980*b*). Speciation in *Macaca*: the view from Sulawesi. In *The macaques: studies in ecology, behavior and evolution* (ed. D. G. Lindburg), pp. 84–124. Van Nostrand Reinhold, New York.

Gruas-Caragnetto, C. (1976). Etude palynologique du Palaeogene du Sud de l'Angleterre. *Cahiers de Micropaleontologie* **1**, 1–49.

Grubb, P. J. (1977). Control of forest growth and distribution on wet tropical mountains: with special reference to mineral nutrition. *Annual Reviews of Ecology and Systematics* **8**, 83–107.

Haffer, J. (In preparation). Quaternary history. In *Biogeography and quaternary history in tropical America* (ed. T. C. Whitmore).

Haile, N. S. (1958). The snakes of Borneo, with a key to the species. *Sarawak Museums Journal* **8**, 743–71.

—— (1971). Quaternary shorelines in West Malaysia and adjacent parts of the Sunda Shelf. *Quaternaria* **15**, 333–43.

—— (1978). Reconnaissance palaeomagnetic results from Sulawesi, Indonesia, and their bearing on palaeogeographic reconstructions. *Tectonophysics* **46**, 77–84.

Hallam, A. (1967). The bearing of certain palaeozoogeographic data on continental drift. *Palaeogeography, Palaeoclimatology and Palaeoecology* **3**, 201–41.

Hamilton, W. (1974). Earthquake map of the Indonesian region. *United States Geological Survey Map* 1–875–C.

—— (1978). Tectonic map of the Indonesian region. *United States Geological Survey Map* 1–875–D.

—— (1979). Tectonics of the Indonesian region. *United States Geological Survey Professional Papers* **1078**, 3481.

Hammen, T. van der (1957). Climatic periodicity and evolution of South American Maestrichtian and Tertiary floras. *Boletin geológico. Instituto geológico Nacional, Colombia, Bogotá* **3**, 59–91.

Hatherton, T. and Dickinson, W. R. (1969). The relationship between andesitic volcanisms and seismicity in Indonesia, the Lesser Antilles and other island arcs. *Journal of Geophysical Research* 74, 5301–10

Hooijer, D. A. (1948). Pleistocene vertebrates from Celebes II. *Testudo margae* n.sp. *Proceedings koninklijke Nederlandse Akademie van Wetenschappen* **51**, 1169–82.

—— (1950). Man and other mammals from Toalian sites in south-western Celebes. *Verhandelingen der koninklijke Nederlandse Akademie van Wetenschappen* **46**, 1–165.

—— (1954). Dentition and skeleton of *Celebochoerus heekereni* Hooijer. *Zoologische Verhandelingen Leiden* **24**, 1–46.

—— (1955). Fossil Proboscidea from the Malay Archipelago and the Punjab. *Zoologische Verhandelingen Leiden* **28**, 1–146.

—— (1957*a*). A *Stegodon* from Flores. *Treubia* **24**, 119–29.

—— (1957*b*). Three new giant prehistoric rats from Flores, Lesser Sunda Islands. *Zoologische Mededeelingen, Leiden* **35**, 229–314.

—— (1960). The Pleistocene vertebrate fauna of Celebes. *Asian Perspectives* **2**, 71–6.

—— (1964). Pleistocene vertebrates from Celebes. XII. Notes on pygmy stegodonts. *Zoologische Mededeelingen, Leiden* **40**, 37–44.

—— (1965). Note on *Coryphomys buhleri* Schaub, a gigantic murine rodent from Timor. *Israel Journal of Zoology* **14**, 128–33.

—— (1967). Indo-Australian insular elephants. *Genetica* **38**, 143–62.

—— (1969*a*). The stegodon from Timor. *Proceedings koninklijke Nederlandse Akademie van Wetenschappen* **B72,** 203–10.

—— (1969*b*). Pleistocene vertebrates from Celebes. XIII. *Sus celebensis* Müller and Schlegel, 1845. *Beaufortia* **16**, 215–18.

—— (1972). *Varanus* (Reptilia Sauria) from the Pleistocene of Timor. *Zoologische Mededeelingen, Leiden* **47** 445–80.

—— (1973). Tenzenschildpadden en dwergolifanten. *Museologia* **1**, 9–14.

—— (1974). *Elephas celebensis* (Hooijer) from the Pleistocene of Java. *Zoologische Mededeelingen, Leiden* **48**, 85–93.

—— (1975). Quaternary mammals west and east of Wallace's line. *Netherlands Journal of Zoology* **25,** 46–56.

Huxley, T. H. (1868). On the classification and distribution of the Alectoromorphae and Heteromorphae. *Proceedings of the Zoological Society of London*, 294–319.

Inger, R. F. (1966). The systematics and zoogeography of the Amphibia of Borneo. *Fieldiana, Zoology* **52**, 1–402.

—— and Chin, P. K. (1962). The freshwater fishes of North Borneo. *Fieldiana, Zoology* **45**, 1–268.

Johnson, B. D., Powell, C. McA., and Veevers, J. J. (1976). Spreading history of the eastern Indian Ocean and Greater India's northward flight from Antarctica and Australia. *Bulletin of the Geological Society of America* **87**, 1560–6.

Johnson, L. A. S. and Briggs, B. G. (1975). On the

Proteaceae, the evolution and classification of a southern family. *Botanical Journal of the Linnean Society* **70**, 88–182.

Johnson, R. W. and Jaques, A. L. (1980). Continent-arc collision and reversal of arc polarity: new interpretations from a critical area. *Tectonophysics* **63**, 111–24.

Kalkman, C. (1955). A plant-geographical analysis of the Lesser Sunda Islands. *Acta Botanica Neerlandica* **4**, 200–25.

Kampen, P. N. van (1923). *The Amphibia of the Indo-Australian archipelago.* Brill, Leiden.

Katili, J. A. (1975). Volcanism and plate tectonics in the Indonesian island arcs. *Tectonophysics* **26**, 165–88.

—— (1978). Past and present geotectonic position of Sulawesi, Indonesia. *Tectonophysics* **45**, 289–322.

Kemp, E. M. and Harris, W. K. (1975). The vegetation of Tertiary islands on the Ninetyeast Ridge. *Nature, London* **258**, 303–7.

Keng, H. (1978*a*). The genus *Phyllocladus. Journal of the Arnold Arboretum* **59**, 249–73.

—— (1978*b*). The delimitation of the genus *Magnolia* (Magnoliaceae). *Garden's Bulletin Singapore* **31**, 127–31.

Kidd, R. B. and Davies, T. A. (1978). Indian Ocean sediment distribution since the Late Jurassic. *Marine Geology* **26**, 49–70.

Kiew, R. (1976). The genus *Iguanura* Bl. (Palmae). *Garden's Bulletin, Singapore* **28**, 191–226.

Klompe, Th, H. F. (1956). The structural importance of the Sula Spur. *Proceedings of the 8th Pacific Science Congress* **11A**, 689–888.

Klootwijk, C. T. and Peirce, J. W. (1979). India's and Australia's pole path since the late Mesozoic and the India-Asia collision. *Nature, London* **282**, 605–7.

Koesoemadinata, R. P. and Pulunggono, A. (1975). Geology of the southern Sunda Shelf in reference to the tectonic framework of Tertiary sedimentary basins of western Indonesia. *Geologi Indonesia* **2**, 1–11.

Kowal, N. E. (1966). Shifting cultivation, fire and pine forests in the Cordillera Central Luzon, Philippines. *Ecological Monographs* **36**, 389–419.

Kundig, E. (1956). Geology and ophiolite problems of east Celebes. *Verhandelingen van het nederlandsch Geologisch-Mijnbouwkundig Genootschap* **16,** 210–35.

Lakhanpal, R. N. (1970). Tertiary floras of India and their bearing on the historical geology of the region. *Taxon* **19**, 675–94.

Lack, D. (1976). *Island biology illustrated by the land birds of Jamaica.* Blackwell, Oxford.

Lam, H. J. (1934). Materials towards a study of the flora of the island of New Guinea. *Blumea* **1**, 115–59.

—— (1938). Studies in phylogeny 2. On the phylogeny of the Malaysian *Burseraceae-Canarieae* in general and of *Haplolobus* in particular. *Blumea* **3**, 126–58.

Lattin, G. de (1967). *Grundriss der Zoogeographie.* Fischer, Jena.

Laurie, E. M. D. and Hill. J. E. (1954). *List of land mammals of New Guinea, Celebes and adjacent islands.* British Museum (Natural History), London.

Lincoln, G. A. (1975). Bird counts either side of Wallace's line. *Journal of Zoology* **177,** 349–61.

Logano-C., G., Hernández, J. I., and Henao-S., J. E. (1979). El genero *Trigonobalanus* Forman en el neotropica I. *Caldasia* **12**, 517–37.

Luyendykt, B. P. (1974). Gondwanaland dispersal and the early formation of the Indian Ocean. In T. A. Davies, B. P. Luyendyk, *et al. Initial reports of the deep sea drilling project* **26** pp. 945–52. US Government Printing Office, Washington DC.

Lyon, M. W. (1913). Treeshrews: an account of the mammalian family Tupaiidae. *Proceedings of the United States National Museum* **45**, 1–186.

Mabberley, D. J. (1979). The species of *Chisocheton* (Meliaceae). *Bulletin of the British Museum (Natural History)* **6**, 301–86.

MacArthur, R. H. and Wilson, E. O. (1967). *The theory of island biogeography.* Princeton University Press.

McDowall, R. M. (1978). Generalised tracks and dispersal in biogeography. *Systematic Zoology* **27**, 88–104.

McGowran, B. (1978). Stratigraphical record of early Tertiary oceanic and continental events in the Indian Ocean region. *Mar. Geol.* **26**, 1–39.

Machin, J. (1971). Plant microfossils from Tertiary deposits of the Isle of Wight. *New Phytologist* **70**, 851–72.

Manabe, S. and Hahn, D. G. (1977). Simulation of the tropical climate of an ice age. *Journal of Geophysical Research* **82**, 3889–911.

Marchant, J. (1916). *Alfred Russel Wallace: letters and reminiscences* (2 vols). Cassell, London.

Matharel, M. de, Klein, G., and Oki, T. (1976). Case history of the Bekapai Field. *Proceedings of the Indonesian Petrological Association (5th Annual Convention, June 1976 Jakarta).*

Mathur, Y. K. (1963). Studies in the fossil microflora of Kutch, India: I. On the microflora and of the Hystrichosphaerids in the gypseous shales (Eocene) of western Kutch. *Proceedings of the Natural Institute of Science India* **29**, 356–71.

Mayr, E. (1944). Wallace's line in the light of recent zoogeographic studies. *Quarterly Review of Biology* **19**, 1–14.

Medway, Lord (1972). The Quaternary mammals of Malesia. In *The Quaternary era in Malesia* (ed. P. S.

Ashton and M. Ashton). Geography Department, University of Hull.
—— (1977). Mammals of Borneo. *Monographs of the Malay branch of the Royal Asiatic Society* **7**.
—— (1979). The Niah excavations and an assessment of the impact of early man on mammals in Borneo. *Asian Perspectives* **20**, 51–69.
Meeuwen, M. S. van, Nooteboom, H. P., and Steenis, C. G. G. J. van (1961). Preliminary revisions of some genera of Malaysian Papilionaceae. *Reinwardtia* **5**, 49–56.
Mertens, R. (1930). Die Amphibien und Reptilien der Inseln Bali, Lombok, Sumbawa und Flores. *Abhandlungen hrsg. von der Senckenbergischen Naturforschenden Gesellschaft*, 115–344.
Mertens, R. (1934). Die Insel-Reptilien. *Zoologica (Stuttgart)* **32**, 1–209.
Misonne, X. (1969). African and Indo-Australian Muridae: evolutionary trends. *Annalen koninklijk Museum voor midden Africa* Ser. 8, *Zoologische* **172**, 1–219.
Mitchell, A. H. G. (1981). Phanerozoic plate boundaries in mainland S.E. Asia, the Himalayas and Tibet. *Journal of the Geological Society of London* **138**, 109–22.
Moore, H. E. Jr. (1957). *Veitchia*. *Gentes Herbarum* **8**, 483–536.
—— (1969*a*). *Satakentia*—a new genus of Palmae-Arecoideae. *Principes* **13**, 3–12.
—— (1969*b*). Pacific Palms II. *Principes* **13**, 67–76.
—— (1970). The genus *Rhopaloblaste* (Palmae). *Principes* **14**, 75–92.
—— (1973*a*). Palms in the tropical forest ecosystems of Africa and South America. In *Tropical forest ecosystems in Africa and South America: a comparative review* (ed. B. J. Meggers, E. S. Ayensu, and W. D. Duckworth). Smithsonian Institution Press, Washington.
—— (1973*b*). The major groups of palms and their distribution. *Gentes Herbarum* **11**, 27–141.
—— (1978). *Tectiphiala*, a new genus of Palmae from Mauritius. *Gentes Herbarum* **11**, 284–90.
Morley, R. J. (1977). Palynology of Tertiary and Quaternary sediments in southeast Asia. *Proceedings Indonesian Petroleum Association 6th Annual Convention*, 255–76.
Muller, J. (1966). Montane pollen from the Tertiary of north-west Borneo. *Blumea* **14**, 231–5.
—— (1968). Palynology of the Fedawan and Plateau sandstone formations (Cretaceous-Eocene) in Sarawak, Malaysia. *Micropalaeontology* **14**, 1–37.
—— (1972). Palynological evidence for change in geomorphology, climate and vegetation in the mid-Pliocene of Malesia. In *The Quarternary era in Malesia* (ed. P. S. Ashton and M. Ashton). Geography Department, University of Hull.
—— (1974). *A comparison of southeast Asian with European fossil angiosperm pollen flores*. Birbal Sahni Institute of Palaeobotany Special Publication No. 1, 49–56.
Müller, S. (1846). Ueber den Charakter der Thierwelt auf den Inseln des Indischen Archipels. *Archiv für Naturgeschiente* **12**, 109–28.
Musser, G. G. (1971). The taxonomic association of *Mus faberi* Jentink with *Rattus xanthurus* (Gray), a species known only from Celebes (Rodentia: Muridae). *Zoologische Mededeelingen, Leiden* **45**, 107–18.
Myers, G. S. (1949). Salt-tolerance of fresh-water fish groups in relation to zoogeographical problems. *Bijdragen tot de Dierkunde* **28**, 315–22.
Nelson, G. J. (1974). Historical biogeography: an alternative formalization. *Systematic Zoology* **23**, 555–8.
Norton, I. O. and Sclater, J. G. (1979). A model for the evolution of the Indian Ocean and the break-up of Gondwanaland. *Journal of Geophysical Research* **84**, 6831–9.
Norvick, M. S. (1979). The tectonic history of the Banda Arcs, eastern Indonesia: a review. *Journal of the Geological Society of London* **136**, 519–27.
Ntima, O. O. (1968). The Araucarias. *Fast growing timber trees of the lowland Tropics* **3**. Commonwealth Forestry Institute, University of Oxford.
Pitman, W. C. III and Talwani, M. (1972). Sea floor spreading in the North Atlantic. *Bulletin of the Geological Society of America* **83**, 619–46.
Powell, D. E. (1976). The geological evolution of the continental margin off northwest Australia. *Australasian Petroleum Exploration Association Journal* **16**, 13–23.
Prance, G. T. (1973). Phytogeographic support for the theory of Pleistocene forest refuges in the Amazon basin based on evidence from distribution patterns in Caryocaraceae, Chrysobalanaceae, Dichapetalaceae and Lecythidaceae. *Acta Amazonica* **3**, 5–28.
Prance, G. T. (ed.) (1981). *Biological diversification in the tropics*. Columbia University Press, New York.
Prance, G. T. Plants. In *Biogeography and Quaternary history in tropical America* (ed. T. C. Whitmore). (In preparation.)
Raven, R. C. (1935). Wallace's line and the distribution of Indo-Australian mammals. *Bulletin of the American Museum of Natural History* **68**, 179–294.
Raven, P. H. (1979). Plate tectonics and southern hemisphere biogeography. In *Tropical Botany* (ed. K. Lasen and L. B. Holm-Nielsen). Academic Press, London.

Raven, P. H. and Axelrod, D. I. (1972). Plate tectonics and Australasian palaeobiogeography. *Science, N. Y.* **176**, 1379–86.

—— and —— (1974). Angiosperm biogeography and past continental movements. *Annals of the Missouri Botanical Garden* **61**, 539–673.

Reid, E. M. and Chandler, M. E. J. (1933). *The London clay flora*. British Museum (Natural History), London.

Rensch, B. (1936). *Die Geschichte des Sundabogens*. Borntraeger, Berlin.

Richards, P. W. (1943). The ecological segregation of the Indo-Malayan and Australian elements in the vegetation of Borneo. *Proceedings of the Linnean Society of London* **154**, 154–6.

Roberts, T. R. (1978). An ichthyological survey of the Fly River in Papua New Guinea. *Smithsonian Contributions to Zoology* **281**, 1–172.

Roeder, D. (1977). Philippine arc system—collision of flipped subduction zones. *Geology* **5**, 203–6.

Rojo, J. P. (1979). Updated enumeration of Philippine dipterocarps. *Sylvatrop. The Philippine Forest Research Journal* **4**, 123–46.

Rosen, D. E. 1975. A vicariance model of Caribbean biogeography. *Systematic Zoology* **24**, 431–64.

Sander, N. J. and Humphrey, W. C. (1975). Tectonic framework of southeast Asia and Australasia: its significance in the occurrence of petroleum. *World Petroleum Congress, Tokyo* PD **7**, 1–25.

Sartono, S. (1979). The discovery of a pygmy stegodon from Sumba, east Indonesia: an announcement. *Modern Quaternary Research in southeast Asia* **5**, 57–63.

Sasajima, S., Nishimura, S., Hirooka, K., Otofuji, Y., van Leeuwen, T., and Hehuwat, F. (1980). Palaeomagnetic studies combined with fission-track datings on the western arc of Sulawesi, east Indonesia. *Tectonophysics* **64**, 163–72.

Schmidt, F. H. and Ferguson, J. H. A. (1951). Rainfall types based on wet and dry period ratios for Indonesia with western New Guinea. *Verhandelingen Djawatan Meteoroligi dan Geofisik Djakarta* **42**.

Schuster, R. M. (1972). Continental movements, Wallace's line and Indo-Malayan Australasian dispersal of land plants, some eclectic concepts. *Botanical Review*, **38**, 3–86.

—— (1976). Plate tectonics and its bearing on the geographical origin and dispersal of angiosperms. In *Origin and early evolution of angiosperms* (ed. C. B. Beak). Columbia University Press, New York.

Sclater, P. L. (1858). On the general geographical distribution of the members of the class Aves. *Journal of the Linnean Society of London* **2**, 130–45.

Scrivenor, J. B. *et al.* (1943). A discussion of the biogeographic division of the Indo-Australian archipelago, with criticism of the Wallace and Weber lines and of any other dividing lines and with an attempt to obtain uniformity in the names used for the divisions. *Proceedings of the Linnean Society of London* **154**, 120–65.

Silas, E. G. (1953). Classification, zoogeography and evolution of the fishes of the cyprinoid families Homalopteridae and Gastromyzonidae. *Record of the Indian Museum* **50**, 173–264.

—— (1958). Study on cyprinid fishes of the oriental genus *Chela* Hamilton. *Journal of the Bombay Natural History Society* **55**, 55–99.

Simpson, G. G. (1977). Too many lines; The limits of the oriental and Australian zoogeographic regions. *Proceedings of the American Philosophical Society* **121**, 102–20.

Sleumer, H. (1955). Proteaceae. *Flora Malesiana* Ser. 1, **5**, 147–206.

—— (1964). Epacridaceae. *Flora Malesiana* Ser. 1, **6**, 422–44.

Smith, A. C. (1943). Taxonomic notes on the Old World Species of Winteraceae. *Journal of the Arnold Arboretum* **24**, 119–64.

—— (1970). The Pacific as a key to flowering plant history. *University of Hawaii, Harold L. Lyon Arboretum Lectures* **1**, 1–28.

Smith, A. G. and Briden, J. C. (1977). *Mesozoic and Cenozoic palaeocontinental maps*. Cambridge University Press, Cambridge.

Smith, A. G., Hurley, A. M., and Briden, J. C. (1980). *Phanerozoic palaeocontinental maps*. Cambridge University Press.

Smith, J. P. (1977). Continental drift and climate. *Nature, London* **266**, 592.

Smythies, B. E. (1968). *Birds of Borneo* (2nd edn). Oliver & Boyd, Edinburgh.

Soepadmo, E. (1972). Fagaceae. *Flora Malesiana* Ser. 1, **7**, 265–403.

South, P. J. (1977). Continental drift changes climate. *Nature, London* **266**, 592.

Steenis, C. G. G. J. van (1934). On the origin of the Malaysian mountain flora, *Bulletin Jardin Botanique Buitenzorg* Ser. 3, **13**, 135–262.

—— (1936). On the origin of the Malaysian mountain flora, 3. Analysis of floristical relationships (first instalment). *Bulletin Jardin Botanique Buitenzorg* Ser. 3, **14**, 56–72.

—— (1950). The delimitation of Malaysia and its main plant geographical divisions. *Flora Malesiana* Ser. 1, **1**, lxx–lxxv.

Steenis, C. G. G. J. van (1957). Outline of vegetation types in Indonesia, and some adjacent regions. *Proceedings of the Pacific Science Congress* **8,** 61–97.

—— (1962). The mountain flora of the Malaysian tropics. *Endeavour* **21**, 183–92.

—— (1965). Plant geography of the mountain flora of Mt. Kinabalu. *Proceedings of the Royal Society* **B 161**, 7–38.

—— (1971). *Nothofagus*, key genus of plant geography in time and space, living and fossil, ecology and phylogeny. *Blumea* **19**, 65–98.

—— (1972*a*). *The mountain flora of Java*. Brill, Leiden.

—— (1972*b*). *Nothofagus*, key genus to plant geography. *Symposium taxonomy, phytogeography and evolution, Manchester 1971*, pp. 275–88. Academic Press, London.

—— (1979). Plant geography of east Malesia. *Botanical Journal of the Linnean Society* **79**, 97–178.

Stein, N. (1978). *Coniferen im Westlichen Malayischen Archipel*. Junk, The Hague.

Stoneley, R. (1974). Evolution of the continental margins bordering a former southern Tethys. In *The geology of continental margins* (ed. C. A. Burk and C. L. Drake). Springer, New York.

Stresemann, E. (1939). Die Vogel von Celebes. *Journal für Ornithologie* **87**, 299–425.

Sufi, S. M. K. (1956). Revision of the oriental fishes of the family Mastacembelidae. *Bulletin of the Raffles Museum* **27**, 93–146.

Sukamto, R. (1975*a*). *Geologic map of Indonesia. Sheet Ujong Pandang 1: 1 000 000*. Geological Survey of Indonesia.

—— (1975*b*). The structure of Sulawesi in the light of plate tectonics. *Regional Conference of Geology and Mineral Resources of Southeast Asia. Association of Indonesian Geologists, Jakerta.*

Takhtajan, A. L. (1969). *Flowering plants: origin and dispersal* (trans. C. Jeffrey). Oliver & Boyd, Edinburgh.

Taylor, E. H. (1968). *The Caecilians of the world: a taxonomic review*. University of Kansas Press, Lawrence.

Thompson, J. E. (1967). A geological history of eastern New Guinea. *Australasian Petroleum Exploration Association Journal* **7**, 83–93.

Traulau, H. (1964). The genus *Nypa* van Wurmb. *Kungliga Svenska vetenskapsakademiens Handlingar* Ser. 4, **10**, 1–29.

Umbgrove, J. H. F. (1949). *Structural history of the East Indies*. Cambridge University Press.

Veevers, J. J. and Heirtzler, J. R. (1974). Tectonic and palaeogeographic synthesis of e.g. 27. *Initial report of deep sea drilling project* **27**, 1049–54.

—— and McElhinny, M. W. (1976). The separation of Australia from other continents. *Earth Sciences Review* **12**, 139–59.

Verstappen, H. Th. (1975). On palaeoclimates and landform development in Malesia. In *Modern Quaternary research in southeast Asia* (ed. G. J. Bartstra and W. A. Casparie). Balkema, Rotterdam.

Vink, W. (1970). The Winteraceae of the Old World. I. *Pseudowintera* and *Drimys*, morphology and taxonomy. *Blumea* **18**, 225–354.

Walker D. (ed.) (1972). *Bridge and barrier: the natural and cultural history of Torres Strait*. Australian National University Research School of Pacific Studies, Department of Biogeography and Geomorphology, Publication BG/3, Canberra.

—— (1981). Speculations on the origins and evolution of Sunda-Sahul rain forests. In G. T. Prance (ed.) *Biological diversification in the tropics*. Columbia University Press, New York.

—— and Flenley, J. R. (1979). Late Quaternary vegetational history of the Enga Province of upland Papua New Guinea. *Philosophical Transactions of the Royal Society* **B 286**, 265–344.

Wallace, A. R. (1859). Letter from Mr Wallace concerning the geographical distribution of birds. *Ibis* **1**, 449–54.

—— (1860*a*). The ornithology of northern Celebes. *Ibis* **2**, 140–7.

—— (1860*b*). On the zoological geography of the Malay archipelago. *Journal of the Linnean Society of London* **14**, 172–84.

—— (1862). List of birds from the Sula Islands (east of Celebes), with descriptions of the new species. *Proceedings of the Zoological Society of London* 333–46.

—— (1863). On the physical geography of the Malay Archipelago. *Journal of the Royal Geographical Society* **33**, 217–34.

—— (1865). On the pigeons of the Malay Archipelago. *Ibis* new series, **1**, 365–400.

—— (1869). *The Malay Archipelago* (2 vols). Macmillan, London.

—— (1876). *The geographical distribution of animals* (2 vols). Macmillan, London.

—— (1880). *Island life*. Macmillan, London.

—— (1910). *The world of life*. Chapman and Hall, London.

Weber, M. (1904). *Die Säugetiere. Einfünrung in die Anatomie und Systematik der Recenten und Fossilen Mammalia*. Fischer, Jena.

Weber, M. and de Beaufort, L. F. (1911). *The fishes of the Indo-Australian Archipelago*. E. J. Brill, Leiden.

Webster, P. J. and Streten, N. A. (1978). Late Quaternary ice age climates of tropical Australasia: interpretations and reconstructions. *Quaternary Research* **10**, 279–309.

Wegener, A. (1924). *The origin of continents and oceans* (3rd edn). Methuen, London.

White, F. (1978). The Guineo-Congolian region and its relationships to other phytochoria. *Bulletin du Jardin Botanique National de Belgique* **49**, 11–55.

Whitmore, T. C. (1975). *Tropical rain forests of the Far East*. Clarendon Press, Oxford.

—— (1977*a*). *Palms of Malaya* (rêvised edn). Oxford University Press, Kuala Lumpur.

—— (1977*b*) A first look at *Agathis*. *Tropical Forestry Papers* **11**.

—— (1980). A monograph of *Agathis*. *Plant systematics and evolution* **135**, 41–69.

—— (ed.) *Biogeography and Quaternary history in Tropical America*. (In preparation.)

—— and Page, C. N. (1980). Evolutionary implications of the distributions and ecology of the tropical conifer *Agathis*. *New Phytologist* **84**, 407–16.

Zaklinskaya, E. D. (1978). Palynological information from late Pliocene–Pleistocene deposits recovered by deep-sea drilling in the region of the island of Timor. *Review of Palaeobotany and Palynology* **26**, 227–41.

Zollinger, H. (1857). Over het begtip enden omvang eener Flora Malesiana. *Natuurkundig Tidschrift voor Nederlandisch-Indië* **13**, 293–322.

INDEX OF ANIMAL AND PLANT NAMES

Page numbers in **bold type** refer to pages on which illustrations occur.

GENERAL INDEX

Page numbers in **bold type** *refer to pages on which illustrations occur.*

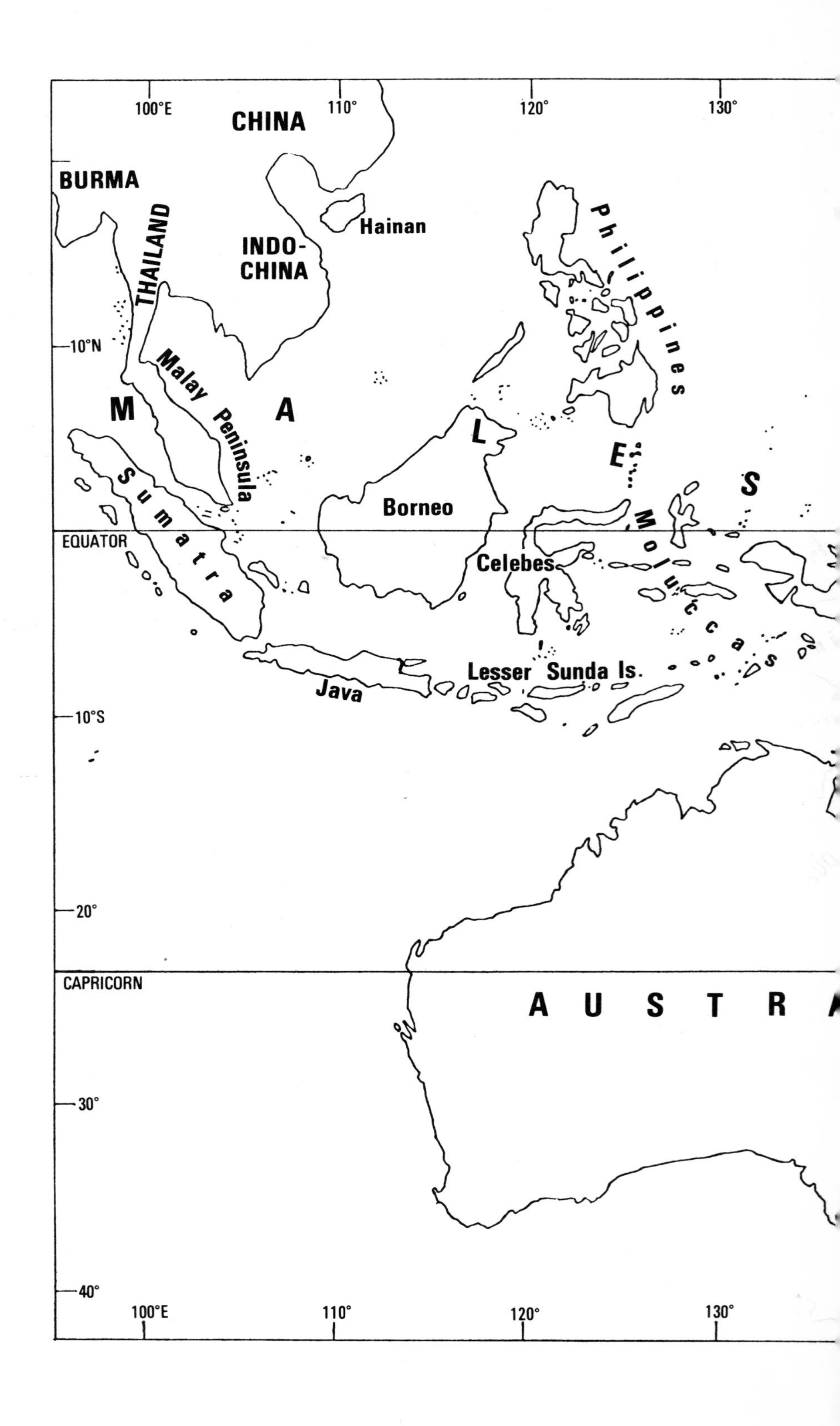

100°E
110°
120°
130°
CHINA
BURMA
THAILAND
Hainan
INDO-
CHINA
Philippines
10°N
Malay Peninsula
M
A
L
E
S
Sumatra
Borneo
EQUATOR
Celebes
Moluccas
Lesser Sunda Is.
Java
10°S
20°
CAPRICORN
AUSTR
30°
40°
100°E
110°
120°
130°